The Brain, God
And
Key Thought Processes

By
Margaret Hardway

The Brain, God, and Key Thought Processes
By Margaret Josephine Hardway

ISBN # 978-0-578-04305-0

Printed In the United States of America
Firestarter Publications
4355 Hwy 43S, Harrison, Arkansas 72601, 870-619-4052
Margaret Hardway
8571 Enault Lane, Garden Grove, California 92841-3270
714-586-0815

Library of Congress Cataloging-in-Publication Data

Scripture Quotations from the Amplified New Testament © Zondervan
Corporation 1975, New King James Version © Thomas Nelson Inc. 1982

CONTENTS

Illustrations:

INTRODUCTION

The Spartan truth is that many need a reprieve from the turbulent load on their minds. Emotions kept in the sea of no return can suddenly resurface with old patterns and cycles that need to be changed! They need to see through the fog of their current lifestyles being so attached to their old ways like barnacles clinging to the hull of the boat.

When I was in my twenties my friends talked me into becoming a licensed scuba diver. The dream was to become the next generation of deep sea adventurers following in Jacque Cousteau's footsteps. On one of our adventures, everyone on the sailboat informed me that I had been chosen for the high honor to scuba dive under the boat and scrub off all the barnacles on the bottom of the boat that were causing it to drag through the ocean when it should glide through channel waters.

Interestingly enough in the deep sea of infinite thoughts we assimilate in our mind, we can have a buildup of so called barnacle-drag or mental cravings that are not in synchronization with how it was designed to function. For instance when you lack proper neurotransmitters, or brain supplementation,

Thought fragmentation or burnout occurs. Your brain was designed to be vibrantly brilliant with all it amazing features of rapid fire. Learn to lead captivity-thoughts-captive instead of the negative thoughts in your mind leading you.

Realign your mind with "Brain-Tuning", which is one of the most amazing tools and techniques that cause immediate transformative breakthrough in many areas of your life. You become the New-Freed-Dream-Leader in understanding how your Bio-Rhythm time clock works with Creator God and your mind. How do we retrain your brain into the best thinking patterns?

What is this grand symphony of a trillion neurons that crosstalk to each other, and what are the unusual encoding sounds in the higher population of neurons that wire and fire together? These are just a few of things we'll cover when we tap into "Old-Faithful" the geyser of your Brain and you're Key-Thought-Processes.

ACKNOWLEDGEMENTS

I want to give thanks to the many supporters beginning with Barbara and Bobby Kenton for all their efforts in the process of releasing this book, and seeing the vision for excellence in our minds. For what Creator God had endowed us with a rich knowledge of understanding higher thought waves in everything he has released within us.

A special thanks to Jeanne Saul for your confidence and belief in this book to persist in helping with your unique company of Firestarter Publications and with your creative efforts on making this dream a reality.

I especially want to give thanks to Tim Asher for your superlative logos of the emerging elements of the design and helping to make the book a success.

CHAPTER 1

EINSTEIN'S MYSTERIOUS BRAIN

The Jewish rooted scientist known as Albert Einstein was catapulted as a javelin from the womb of time, to fathom the guarded secrets of the universe. Inside his mind between the valleys and fissures, the waves and wrinkling, lay the epic panorama of universal mysteries. Albert was determined to conquer Mount Everest of the science world. He was a quintessential celebrity with his bolts-of-brilliance, which the world heralded by rolling out the red carpet to applaud his endeavors with his landmark scientific discoveries.

He unmasked many secret origins of the universe, explaining the mysteries of nature, space, and time, with his mathematical equations. He stated that "Space is three dimensional cushions sagging beneath bowling balls of heavenly bodies." He loved the arts and mastered playing the music of Mozart. This eminent, quotable prophet was at one time offered the presidency of Israel. His roots were grafted into the God of Israel, as a child trained in Hebrew school.

Einstein transitioned his brain to champion the rights of the underdog, as a David fighting against Goliath. According

To Dr. Marian Diamond, former head of Hall of Science at UC Berkeley in the mid-1980 stated that the brain decides its own future. Dr. Diamond quoted the fact "That the nervous system possesses not just a morning of plasticity, but an afternoon, and evening as well". An older healthy brain supported by an active lifestyle, with continuous mental-stimulation, enhanced by a nutrient enriching brain diet, is able to function as well as a young healthy brain.

Tom Harvey a pathologist was entrusted with Einstein's brain first. He dissected it into 240 pieces and placed it into formaldehyde. Many Years later Dr. Diamond was randomly chosen to dissect and study Albert Einstein's brain along with other scientists. Why are genius brains different from normal people's brains? This question was posed to Dr. Diamond as she unraveled a mysterious clue that even Einstein himself once remarked, "That when he was in deep thought, words played no part in his mental dialogues, instead his thoughts were a combination of certain signs and more or less clear images."

This reminds me of the passage in *Acts 2:17 "It shall come to pass in the last days, says God, I will pour out of my spirit upon all flesh: and your sons and daughters shall prophesy, and your young men will see visions and your old men dream dreams."* Fulfilling what Einstein saw in conceptual signs and visions...in other words, his most productive thoughts were primarily deeply abstract, his epiphany-revelations of the science cosmos was epoch. He proclaimed invincible writings on traveling at the "speed-of-thought" therefore; Dr. Diamond decided to examine carefully

2

an area of Einstein's brain that was most intricately involved with imagery and abstract reasoning, the Superior Prefrontal and Inferior Parietal lobes. When Dr. Diamond studied Einstein's brain; she compared it with eleven other human brains, which were harvested from intellectually average men who had died at the same age of 76 years old as Einstein.

What Dr. Diamond discovered was there was no discernible physical difference in Einstein's brain and the other brains-with one notable and exciting exception. Had the God of Israel inserted a "Mysterious-Genius-Gene" into this Jewish man to unfurl the mysteries of the universe? Indeed he was head and shoulders above his peers, with his scaffolding intelligence, as Einstein said, "That he wanted to unravel "origins-of-the-universe!"

The conclusion that Dr. Diamond assessed was that Einstein had an increased concentration of a certain type of cell in one unique area of the brain which was known as area 39, and this area is located in the upper back part of the brain known as the Parietal Lobe.

This area was enriched as a highly developed site which included a special type of cell that clustered in superabundance in Einstein's brain known as the "Glial cell." Glial cells are very common in the brain, but they are, labeled in effect, as the brain's "housekeeping" cells and their primary job is to maintain and support the metabolism of the thinking neurons.

An example of this is compared to our Vice President supporting the President. This meant Einstein's thinking cells needed a great deal of metabolic support. Why would they need this much help? Because they were doing volumes of work, and constantly involved in Einstein's deep and hard thinking. There

Is a revelation of this penned in ***Psalms: 92:5 "Oh Lord how great is your works and your "thoughts" are very deep!"*** The abundance of Glial cells had produced an enlargement in Einstein's brain known as area 39. Obviously this German-Jewish man was born with blessings of an excellent brain and overflowing fluid intelligence. When Einstein passed in 1955 he donated all his documents, papers, and Thesis writings to the prestigious {Jerusalem-Hebrew-University} as one of the founders. So when they transferred all his estate documents, they were guarded and escorted by Israeli police into a safe vault.

Einstein's brain was eventually returned to Princeton hospital where another new study was performed by Dr. Elliot Krauss. He concluded that the part of his brain known as "Sylvian Fissure" located in the parietal lobes was structurally different than a normal brain. Since that area was used in mathematics and spatial reasoning he noted this area was 15% wider.

An example of this is stated in the *Ephesians 3:9 "And to make all men see what is the fellowship of the "mystery," which from the beginning of the world has been hid in God, who created all things by Jesus Christ."*

Fluid intelligence is the measure of how streamlined our brain operates in its multi-functions. Einstein's God-given fluid intelligence was a prototype of a master-mind genius, not the amount of facts compiled into it. He had enlarged the most important part of his brain by mentally exercising it to the maximum possible degree. Einstein was an elite-mental-athlete who trained hard all his life. Whether we are young or old, we can continue to learn.

These were the key vital thought processes that enabled him to revolutionize science. He showed that "absolute-time" had to be replaced by a new absolute; "The speed-of-light" which produced the decreed equivalence of mass and energy which led to his famous discovery of $E=mc2$. Amazingly Creator God speaks of this light in *Psalms: 104: 2 "Who covers himself with "light" as with a garment, who stretches out the heavens like a curtain".*

The Hebrew writer portrays Creator God as magnified "light" as his covering! This makes me think of Satellites and there uncanny ability to retrieve images from the ocean. In the sea there are various pockets of light that spread as wide as 100 miles. These unusual shafts of light contain a luminescence which is known as the "Glowing Milky Seas" which are filled with myriads of tiny bacteria worm called infusorians.

These tiny creatures glow with a continuous "milky light" causing sparkling waves behind boats as they glide through the ocean. I personally experienced this as I was surfing on California coast one summer night when I noticed a phenomenal black-light glow in the ocean which appeared in the waves of red tide, it was fantastic to see!

Amazingly Einstein challenged the Wave Theory of Light, by stating that it consists of collected-particles. This won him the Nobel Peace Prize in 1921 for his work on {photo-electric-effect}. In this he suggested that "Light waves were stream of Quantum-Packets called "Photons" thus intensity of light is driven by a number of photons and energy light is determined by photon energy, which later opened the door for "Quantum-Physics" Einstein's doctoral-mindset assimilated ideas sputtering around as calculated molecules of dimensions; suspended in stationary liquids.

A simple example of this is displayed in my friend's teenager rooms that have lava lamps, when they plug in their lights they glow with floating stationary liquids tumbling around in the lamps! I thought in a sense we are small quantum particles of God's light of his calculated molecules, within planet earth.

Einstein helped to advance our modern technology that we enjoy to day such as radios, televisions, remote controls, and transportation aspects of his energy theories.

He once went to the Grand Canyon in 1931 and visited with Native American Hopi tribe, they honored him by crowning him with their native headdress! Jew and the Native American alike were celebrating the linking of two tribes! He mastered his brain and was a Hailed-Hebrew-Genius of extraordinary brilliance!

Psalm 139:14-16 "I will praise you for I am fearfully wonderfully made. Your eyes did see my substance yet being un-perfect, and in your book all my members were written which in continuance you made, when as yet there was none of them." In the beginning of a tiny nerve cell the embryo channels the first branch outwards, to overcome ignorance gathering knowledge with increasing creativity. Then we become a little more idealistic, generous and altruistic, with microscopic dendrites-tree-branches that increase memory.

The overwhelming evidence that factors emotional-intelligence is often put on the balance scales with intellectual-intelligence. Our neurons circle around a high-wire-neural-circuitry, similar to the epic adventure of the tales of the Knights of the Round Table with King Arthur. Or consider Christ at the Last Supper with his Round Table of the chosen twelve Apostles.

CHAPTER 2

PARADISE BRAINS

Your brain is on the voyage of life like a sapphire sea teaming with marine-life filled with billions and billions of organisms. If you consider starfish in the sea, with its uncanny ability to regenerate new life as in one of their arms being cut off, they will grow a new one. The same pattern holds true in our minds that we can also recoup lost neural networks which will grow and expand with the plasticity of an increased new neural army equipped with new thought patterns...."Voila"!
These vital information circuits are ever-changing their communications. That is why teenagers think they can take on the world, because they have over a trillion bursting healthy neurons and synapses firing together actively!

Tiny powerhouses called "Mitochondria," are tiny bean shaped compartments that supply enzymes essential for energy and are considered miniature fountains of youth! They contain a whopping bundle of over 1200 protein cells! There is an estimated 50 diseases associated with defects that result in malfunctioning of these tiny microscopic powerhouses. Mitochondria have their own tiny genome, within a set of chromosomes. They are their own generators with an attached power source.

The press headlines blared "President Obama went to Hawaii for his Christmas vacation, a sudden storm erupted that knocked out his generator so he didn't have power for 11 hours on the Hawaiian island of Oahu". Mitochondria are DNA inherited from the mother in maternal inheritance. Because it's matrilineal ancestor, humans have coined the phrase "Mitochondrial-Eve" from the beginnings of Creation. *Genesis: 3:20 "And Adam called his wife's name Eve; because she was the mother of all living".*

According to Jeff Victoroff M.D. in his book "Saving Your Brain" when we get set in our own ways at the expense of new learning, a 3rd aging occurs in our Mitochondria DNA similar to throwing a monkey wrench into our neural engines. They work less efficiently, are more worrisome and making less energy. Like a creaky old car engine dumping fumes out of the back of a misfiring tailpipe, spewing out free radicals which rip neurons to shreds. In the book of *Philippians: 4:6 "Be careful for nothing but by prayer and supplication with thanksgiving let your requests be made known to God. And the peace of God which passes all understanding shall keep your hearts and "minds" through Christ Jesus".*

Through meditation or prayer it sets your minds and thoughts on things above, up and away from the wrecking-ball mindset in life. In relation to *Proverbs: 23:7 "For as a man "thinks" in his heart so is he".*

We can make fine tuning adjustments described in *Philippians: 4:8 "Whatsoever things are honest, whatsoever things are just, whatsoever things are pure, whatsoever things are of a good report; if there be any virtue, and if there be praise, think-on-these-things."*

Through these windows of communication it is imperative to have the Clarion-Trumpet call of Creator's thoughts in our minds. According to Michael Moizen M.D, your brain is like a city's power plant with its pulsing electrical grid that transmits signals to homes and businesses. But when a hurricane storm knocks out power lines in a blackout; via life's storms impact your brain's power lines get knocked out, and you may experience a disconnection in your synapses which would cause a blackout-thought fragmentation. "Hmmm where are my car-keys"? "Did I leave them at the university"? "Why can't I think straight"?

In fact you could compare this to a car alternator that goes out on you and begins to pull life from the battery. This slowly draws strength from your headlights as they fade into dim because of a low battery. Some people cruise along at snail-speed on life's highway, in a comatose state, like the fairy tale of the Hare and the Tortoise. Our brains were designed for momentum instead. As explained in *Ephesians 3:20 "Now unto him who is able to do exceedingly abundantly above all that we ask or "think" according to the power that works in you."*

This is a metaphor of how Creator God works his power-plant-mind in us. So we can perceive how great the almighty creator God thoughts are towards us! I stepped through the doors of 24 hour fitness club to workout and they had posters plastered all over the walls with bold captions displaying a woman swimming in the lap pool that stated "she had just recovered from brain surgery."

She was assured that her brain was on the highway to healing as she retrained in a gym environment, to improve her brain state of conditioning, as well as her body. It takes time for a perception to reach the brain, and canvass new routines.

The Parietal lobe is the area of the brain that processes pain and pleasure. This area can fool us into thinking "Oh no that is too painful to change in that area, as my friend Joy Asher plainly reveals in her writings on "Issues of the Heart". All we need to do is think critically in these terms by asking Creator God a simple question. "How can I change this area in my life to overcome the hurdles that are set before me"?

Just like the woman at 24 Hour Fitness Club did! This way the brain will not keep you from change, but you'll simply shift into making the adjustments necessary. Evidence clearly reveals that extended exercise is fantastic for the brain because it stimulates memory and brings physical changes in the brain structure.

Like a voting booth where decisions are obeyed, vetoed or postponed. How is this possible in *Ephesians 4:23 "And be renewed in the spirit of your mind."* Because of our infinite wisdom of creativity we don't need brain surgery but instead we can impute learning a new operating system on our computer. This new system has improved commands, formats, different upgrades, puzzle languages. New ideas and changes such as young entrepreneur Mark Zuckerberg had when he broke out of his college box five years ago as creator and founder of Face book with 175 million users accessing it.

An interesting study was done on a group of pigeons who were trained to distinguish between Picasso paintings and Monet paintings. Amazingly this showed the alignment that will work with the power of change even in the most primitive bird-brain Neocortex. In the European Educational system they routinely teach students two or more languages. In light of this consider this passage in *Philippians 4:13 "I can do all things through Christ who strengthens me."*

In the industry of heart-pacemakers in their endeavors to preserve the heart, there is hope on the horizon. Scientists are at the threshold of figuring out a way to add a pacemaker in the brain with "Deep-Brain-Stimulation" which would help Parkinson's disease according to Rob-Stein's article in the Washington-Post. In our brain's hub we're enabled to open hidden doors with keys reflecting a larger framework.

When we think in a new paradigm shift fogginess or mental dullness can lift off our mind, and we're empowered with squeegee clear windows of thinking. Even in a "Brain Spa" our mind visits happy places through visualization. Some movie stars as Goldie Hawn, Jack Nicholson, and Mel Gibson, have gone to Palm Desert, Hot Springs California. Here they have retreats with Native-American-massages complete with rain-sticks, and Native American drumbeats pounded over you. Imagine sinking into the luxurious fresh eucalyptus scrubs, home grown herbs with warm oil, foot, and head massage, mineral water showers, and aromatherapy.

It is amazing to know that you have been created to distinguish 10,000 odors. Fragrant smells can embellish your thoughts and form the most essential form of human congress. Our nostrils create smell maps which are the crème de la crème forming a euphoric place in the Brain Spa. Unfortunately, stressors that dominate our society can shift our thought processes into "Fight or Flight."

This can cause a sinking feeling like quicksand in the economic recession of our mind's paradigm. But wouldn't you love to obtain the bounty that is yours for the asking described in *Proverbs 4:7-9 Wisdom is the principal thing, therefore get wisdom: and with all your getting get understanding. Exalt her, and she shall promote you. She shall bring you to honor, when you embrace her. She shall give to your "Head" an ornament of grace: a crown of glory shall she deliver to you"!*

Thus we can wade into streams of crystal clear rivers as natural as a fish swimming in their environments. My sister and her husband go mountain biking in the pristine Mammoth Mountains in California. They like to fly-fish to catch their fresh-fish-tacos in the raging white water rapids, hauling their Coleman stove and tent into the campsites.

For my birthday I splurged at the billion dollar art museum called J Paul Getty in Los Angeles California, as I gazed upon masterful pieces of painted artworks from different dynasties to the Renaissance period. My mind would wander through a forest of various artists such as Michelangelo, Renoir, Picasso, or Monet's Water Lilies. Lost in my thoughts I was transformed into Renaissance creativity.

I am a Master Painter, and have spent endless hours creating, rich colors of hues and tones on my eclectic collection of various splattered artist palettes, overlapping, blending abstract textures into landscapes creating a silhouette of unusual colors. In the same fashion the brain will take on a silhouette change of overlapping into new ideas to color its world. In the developing of the brain it usually will have the rhetoric changes occur in the back area and progress to the front. The expansion of the brain that integrates with various areas will directly contact with the imagery to color its environment and affect the five senses.

CHAPTER 3

BRAIN'S INTERNAL MAPS

Traveling across America is a lot of fun as long as you have maps for your compass. When I moved across the nation from California to Florida, I signed up for Triple AAA automobile insurance, and they gave me a Ticker-Trek-Map, to navigate a daily route. Eureka! I felt as if I had become one with the early pioneers that trail-blazed the rugged terrain. America's landmarks crowned the landscapes, jutting upwards as towering cathedrals along the Grand Canyon, Yellowstone Park, and New Mexico. The "Eye-Popping" treasure-troves were burrowed in towns along the historic roads, especially the famous route 66.

Nowadays we flip on our GPS mapping systems and it plugs us into our destination. In Neuroscience they have coined the phrase "Dendrites." This is the Greek word for trees or "brain-trees" which are our brain's vital connectors. These contain our internal mapping systems of information, for our thoughts to process memories. Dendrites are of epic importance in brain functions. In each neuron there is a tiny gap called "Synapse" which is the connecting-bridge for your thoughts and memories to swim across into neurotransmitters, like a big bridge transporting you from one place to another.

Thus the more brain trees-and-synapse-connections you have the smarter you are. They are a type of antennae or mapping system that pulse information into our neuron-network-wiring. An example of this is found in the multitudes of satellites hovering above the earth with their mapping systems. The shape of brain trees appears as small trees in winter with no leaves on them. They will ferry across synapses, or little tiny bridges, that carry vital information to the right and left hemispheres of the brain.

Science has discovered almost twenty thousand brain trees will fit on a pinhead! These brain trees are carriers of blood supplies and memory. They sprout as a vast forest of branches across the vast network of neuron-cell-bodies by the tens of thousands like fireworks-exploding into shaped chandeliers. I think of *Psalm 1:3 "He shall be like a tree planted by the rivers of water, that brings forth his fruit in his season; his leaf also shall not wither; and whatsoever he does shall prosper".* We are God's "brain trees" in our blue planet spreading the branches of Creator's thoughts into a place of his kingdom come in the earth.

Each side of the brain has ingeniously separated blood supplies that secrete a chemical called "Neurotrophins" or Nerve-Growth-Factor {NGF}. This works as a type of miracle-grow like fertilizer to your brain. There are an estimated 100 billion neurons or nerve cells which only make up one tenth of the other nine tenths of our cells which are called "Glial." These are multiplied ten-fold over the 100 billion neurons.

A simple comparison of this is if you imagine the brain as a chocolate chip cookie and the neurons were chocolate chips. Glial would be the cookie dough that fills all the space and ensures that the chips are suspended in the appropriate locations. Glial means glue in Greek giving the impression that the main function of these cells is to keep our brains from running out our ears! Glial are supporting cells, and also the housekeepers to neuron thinking cells. Creator God is also a type and shadow of Glial cells and neurons running the systems of our brain-portals! *Colossians: 2:9-10 For in him dwells all the fullness of the Godhead bodily and you are complete in him which is the "Head" of all principality and power.*

Each neuron has a cell body where genes are stored, like your kitchen pantry stocked with food. These contain long extensions called axons; they send messages thru tree-branching-dendrites, in appearance as branches on trees which carry nutrients. They have tiny tubes called microtubules like a subway you would ride in New York. In our complex brain there is calculated at least ten different levels of intensity, which is the total number of brain-states. That's greater than the total number of atoms in our universe. There are incredible brain-states that almighty Creator God contains penned in *Psalms 139:17-18 "How precious also are your thoughts unto me, Oh God! How great is the sum of them! If I should count them; they are more in number than the sand: when I awake, I am still with you".*

The brains "internal maps" have specialized neurons to help cells see vertical or horizontal lines, in receptive fields, this ability allows your eyes to focus on far off objects. Scientists discovered Magnetite Crystals or microscopic magnets in birds brains. These crystal magnets enabled them to read the earth's magnetic field when flying above it, acting as a honing device. The next time you're looking through binoculars and see Canadian geese flying back home to their place of origin, over thousands of miles you'll know the driving source in their feather brains.

The same exists for Loggerhead turtles that have Magnetite crystals in their brains and can navigate the oceans electromagnetic field, up to 8000 miles in warm currents of the Atlantic Ocean. The human brain maps are like a flight of stairs or rungs on a ladder that specialize in color, and motion as an artist would determine their palette of colors, or a movie producer making a motion picture. Their mapped information circuits travel through various networks in the brain, where intricate connections finalize the act of location power.

There is another mapping system that speaks about the mind in *2 Corinthians: 4:3-4 "But if our gospel be hid to them that are lost in whom the God of this world has blinded the "minds" of them which believe not, lest the light of the glorious gospel of Christ, who is the image of God, should shine unto them."*

Entire nations and people groups have there intellects smoke screened, in their ways of thinking. They cannot see truth, as expressed even in *2 Corinthians 3:13-14 "And not as Moses which put a veil over his face, that the children of Israel could not steadfastly look to the end of that which is abolished but their "minds" were blinded, for until this day remains the same veil un-taken away in the reading of the old testament, which veil is done away in Christ"*. In certain places horses are used to pull wagons or carriages behind them, and have blinders on the side of their eyes, so they could only see straight ahead in front of them, not with peripheral vision, to keep them on the road.

CHAPTER 4

BRAIN TUNING

Brain tuning helps your mind to be balanced, since we walk the tightrope high-wire of life with a safety net like trapeze artists use. In brain-tuning alignment, there is an unraveling of thought processes, to enhance clarity, and reveal focused creative empowerment thoughts. One might compare this to a scuba diver plunging into the deep ocean caverns with a scuba light searching for a pearl of great price hidden in your mind. Some people have a safety mechanism set up in their minds to protect them as they live in a bubble, while others withdraw into a shell, to escape their circumstances.

Many have cruised through life's voyage and settled their mindset on a sinking titanic ship, or a victim mentality of Johnny comes lately, "help come rescue me"! Then we see another group of people who have crossed over on a ship called the "Love Boat" reaching out to others on their journey across life's oceans. When the hurricane Katrina slammed into New Orleans we ventured over to the ghost town city that was vacant for months, and helped with Fema in portable tents, food and clothing distribution relief. We stayed at our friend's, Hy's Bed and Breakfast whose home was spared through the storm.

He was a Navy Seal that worked with Mother Teresa in years past. He told us he stayed behind and sent his wife with the evacuation, and faced the storm. He said the tremendous flooding, and winds caused motor boats to float down his street past his house which he used to rescue many people during the storm. He truly was an unsung Hero! Amplified in an important passage in *John: 10:10 "I came that you might have life and have it more abundantly"!* Brain tuning could be compared to when we would tune a musical instrument with musical tuning forks. Remember in the earlier Chapter we referred to two tuning forks, when one sounded, the other fork resonated with the same sounds.

The same in your brain, when sound enters the ears which act as gatekeepers then it will hear the new-frequencies that are resonating in it and open up a new pathway across your neural-network of your brain. An example of this is when a raging river slides down a new pathway after heavy rains, and creates a new course through a mountainside. To experience this phenomenon it is relatively quite easy.

I have released it over many people's brains virtually to see their thought patterns change, and shift out of their old paradigm boxed in mindsets. Literally regenerating any "Lost Thoughts" and long forgotten memories, that might have been disconnected by "thought-fragmentation." There was a popular T.V. Show in the 60's called "Lost-In-Space" which was about the Robinson family that lived in outer space.

They had a peculiar robot which resembled a huge slinky and when trouble or an alien was trying to take over their Jupiter 2 spaceship he would wildly swing his huge-slinky arms yelling "Danger!" Danger"! I think sometimes our brain is

"Lost in its own Space" and we aren't even aware of it! It is set off with an alarm response in a frazzled-frenzy yelling "Danger-Danger!" When I release Brain Tuning it stabilizes your mind and connects you to your pathway of healing, destiny, purpose, and identity. This encompasses a shift in your neural-circuitry thoughts to align with reconnect into proper alignment. Then your brain charges your entire body on a road to recovery, and reconciliation with Creator God, As well as bringing healing to individuals in their thought-processes because in your brain you hold a galaxy of memories and emotions that can become veiled, clogged, or covered up. Sort of like an attic with cobwebs in it after a period of time. So through Brain Tuning we do spring cleaning, which allows vital nerves to increase and connect where they might have been fragmented, through our myriad of roller-coaster emotions, neglect, stress, unpleasant memories, or poor brain diet.

In Neuroscience this would be called Realigning-Your-Neural-Network, or rewiring the hard-wire systems in your mind. We'll say it's an extreme brain makeover! When I begin the brain-tuning process, it begins to synergize and integrate the 4 lobes of the brain, the Frontal lobe, Parietal-lobe, Temporal lobe and the Occipital lobe.

We begin the session by speaking and decreeing to each individual part of the brain lobes I just named. First of all I place my hands on the head and verbalize Creator God's blessing over each part of the brain to come into Resurrection Power with the same power that raised Jesus brain from the dead. Then I align your mind with the "mind" of Christ. Then for all fragmented thinking in the neurons-brain circuitry to fire with the Synapses, and the Dendrites {brain-trees} to connect! Next I command all the tiny parts in the neuron itself to fire and connect which are summarized in this order:

1 Brain Trees-fire
2 Neurons-fire
3 Axons-fire
4 Myelin-Sheath-fire
5 Mitochondria-fire
6 Neurotransmitters-fire
7 Synapses-fire

You're probably wondering why I am commanding the neuron components itself to fire. The reasoning behind it is if your brain trees aren't connecting into the synaptic cleft per say which is your entry way into the neuron then thought fragmentation will settle in because each brain tree carries memory and blood supplies. An illustration of this is like jumper cables connected to a car battery that has charged electrical currents to jump start the car! "Vroom!" In the same way your brain will respond as neurons rush at lightning speed! It has quickly been aligned, onto a superhighway of brain-wave-vibration.

26

Next we proceed to speak the same declarations with Resurrection Power for an increase of new thoughts from Creator God over the:

1 Vagus
2 Amygdala
3 Hypothalamus
4 Thalamus
5 Hippocampus
6 Corpus Callosum

Also included in the tuning for the two hemispheres of {bundles-neurons-axons} to connect and cross-talk from the right brain to the left brain and virtually repeat the same process. I personally say to the axons and neurons "move at the speed-of-light," to connect right and left brains and to be charged-and-fire properly, with "Resurrection Power"! If there is any fragmented thinking to be integrated into the brain trees in Creator Jeshua's name, then to speak and declare and decree to have "Resurrection Power" with the same power that raised Christ from the dead. Commanding the Synapses to connect and fire into integration and as I place my hands on the head for healing.

1 Medulla
2 Brain Stem
3 Pons
4 Cerebellum

This is where I repeat an affirmative decree for all parts of your mind to become strengthened and become whole. The amazing change is stunning and transformative, and immediate for most. It brings a metamorphosis-transition to every vital brain part that you decree this over.

An amazing thing happened from a woman that owned a business called Divine-Release-Spa. I carefully laid my hands on her head and declared over every part of her brains to have an alignment with the mind of Christ to be within her, with the same power that raised Jesus brain from the dead and to have resurrection power in Jesus name. BJ said she didn't feel anything, but for the next 3 days she was at a conference listening to a keynote speaker, and said she was able to recall what was said, and remember everything while she organized her written notes.

This was a tremendous breakthrough for the first time in BJ's life, which she was never able to do before; because in times past she had a mental block and couldn't assimilate her thoughts, and had to write down everything that was taught in any classroom or conferences in order to remember it. Then she told me that the shape of her skull had literally changed. She believed it was a true miracle from Creator God! Astonishingly the breakthrough for her was a miraculous change. Now BJ was enabled to earmark her memories with focused clarity.

Another man told me before we began, that he confided that he felt his brain was numb from an emotional relationship that had gone sour. So when I did the brain-tuning, over his head I decreed Resurrection Power in Jesus name over all his

brain parts in detail, suddenly a burning sensation welled up over his brain, and he said he felt like his whole head was a "lit-up-light-bulb!" With an incredible razor sharp clarity, the heaviness of the mental fog had lifted miraculously from him. He now encountered a sense of newness with the depression vanquished"!

Another woman and her husband said she saw blue "Electric-Aura-Bolts" coming upon her brain, and had all these Incredible ideas come to her for her business. I joked and said it was the Blue-Light-Special.

One of my close friends that live in another state asked for me to do this over the phone, after us Brain-Tuned she said, she could not sleep that night, because she was so amped with new creative ideas flowing in her mind all night long. So she went into her 16 year old son's room and placed her hands on his head and declared brain tuning over her son's brain, the same thing I did over hers, since he was a recluse that stayed in his room all the time.

After she started to declare "Resurrection Power" in Jesus name over all his brain parts, she said the change was remarkable he ventured outside the next day and built a skateboard ramp, and reconnected with all his teenage friends that he was isolated from.

Another friend of mind in Flagstaff Arizona had gone through a lot of rejection issues and emotional scarring in her mind. So I released the decree over her Amygdala and Hippocampus, and she said she felt her brain increased as a tent over her mind and felt instant euphoric peace and relief.

One of the most amazing changes was on a homeless man that was at the Dana Point Christmas boat parade, in California. I was singing Opera-Christmas-carols and he approached me and said he was homeless and was digging for the left over cans that people discarded and told me he had a brain tumor.

I placed my hand on his head where the brain tumor was and declared the "Resurrection Power" in Christ name over his brain. He burst into a startled cry where he said his brain tumor was on his head, and felt a sensation of a "tingling-burning-fire" especially where the brain tumor was, this was truly a remarkable miracle! I felt we were truly rebooting the brain as penned in *Ephesians: 4:15 "But speaking the truth in love may grow up into him in all things which is the "head" even Christ"*.

It is a mystery how our brain coordinates it's intelligence-modeling so quickly. There are slow speeds called "Spikes," not your favorite "spiked-hairdo" though that's what they look like. They jet along, traveling at one foot per second in axons with insulating sheathing called Myelin. Myelin is a layer formed around nerves to allow rapid-speed-impulses in brain nerve cells. Myelin sheathing looks like the plastic that coats any electrical wiring. Myelin in your brain is very important but can be compared to one-hundredth millionth the speed-of-signal-transmission in digital computers.

Yet you can recognize a friend instantly while digital computers are slow and unsuccessful at face recognition, how can our brain operate so quickly? Because it's a "Parallel Processor" running many operations at once, brains are incredibly fast. Sounds like the average American multi-tasking that we daily do, with our mind's ability to rapidly synthesize and parallel process together.

The integration of visual imagery in our Brain-world springs from a complex labyrinthine network, resembling rows of loopy electric telephone wires on poles, probably in part because it's easier to think of our magnificent brain as orderly-assembly-lines than dynamic networks. Isn't that how God is in his word of **Psalms: 139:17 "How precious, also are your "thoughts" unto me, O God! They are more in number than the sand."**

The infinite ways of God's thinking is compared to parallel processors running many operations at once. There is an estimated 6.5 billion people in our earth with almost a trillion neurons operating at once in each person, not to mention creation as well. *1 Corinthians: 2:16 "For who has known the "mind" of the Lord that he may instruct him? But we have the "mind" of Christ."* As we become armed with wisdom and knowledge to bring the mind journey back to its healthful Creator makeover. *1Corinthians: 1:27 "For God has chosen the foolish things of this world to confound the wise."*

Over the counter there is a barrage of supplements for brain-diet-nutrition, such as Omega-3 brain foods, which helps to retain proper memory. Involvement in changing your diet to living foods also is a plus for overall optimum health. One summer I had the luxury in Texas to grow my own Eden-organic-garden full of nutrient-mineral packed vegetables. What a rare treat to have fresh living foods daily on my dinner table harvested from my garden. Though I felt more like the cartoon character Elmer-Fudd chasing hordes of "waskily-wabbits" out of my garden!

There are so many supplements that you can take such as natural enzymes that aids in memory. Here is a list that will help your metabolism to increase your neurotransmitters. These are vitamins that support learning, memory, thinking for short term brain-boost, and reducing memory loss.

- OMEGA 3 Found in fresh fish
- GINKGO BILOBA
- HUPERZINE A
- GINSENG
- CHOLINE
- LECITHIN
- B12
- B6
- B5
- VITAMIN C
- LION'S MANE
- GLUTAMINE

CHAPTER 5

THE VOYAGE OF HIGH RISE THINKING

How is your brain fed? How does it travel into different mind journey's everyday? Better yet, how does it function as a power plant with power grid's glowing and growing within it on a daily basis-very similar to a green house effect? These are just a few of the questions we'll probe into regarding the greatest living organism ever created-a Sherlock Holmes type of investigation into the mysterious-wheels that churn and turn this infinite newly chartered territory. You will be fascinated as you read and find yourself equipped to uncover who you were created to be in your own mind. Your brain could be compared to a high sultan wearing a Royal Crown of knowledge positioned within his palatial splendor over his palatial palace of your life!

Proverbs: 25:2 states "It is the glory of God to conceal a thing: but the honor of kings is to search out a matter". The circuitry within your mind's territorial boundaries holds a treasure of hidden keystones. That unlock superlative mysteries to who you are and what you were created to be for your identity, purpose and destiny.

Seasons change with golden leaves falling from different branches that ebb from trees. Each leaf can be compared to a collective thought flowing down the river of life. So it is in our network of higher-deep-thinking, where we make unique paradigm shifts well it is because in *Proverbs: 23:7 states "As a person thinks in his heart, so are they"*. Our mind's multi-chambered knowledge is compared to portable wealth!

Brain specialists zealously advocate the daily learning of new skills. Our brain has a deeply embedded ability to repair itself. When we choose to learn by embarking upon new tasks that break a set routine our brain cells generate new labyrinths of thought processes.

In this three pound mass of gray and white matter called the brain, we have close to a trillion nerve cells all inter-relating at the speed of thought. This is comparable to the multitude of stars that exist in the Milky Way. New questions cause a hothouse-of-creativity to arise within our brain's ecosystem. When we function on prior data information and patterns we cause our brain to stagnate.

Once our mind fields embrace change, a revolutionary awakening is created that flourishes into a harvest beyond fenced perimeters-challenging creative input together and connecting separate information networks within the very sphere of our brain. As one epic idea transcends from one generation to another our mind has the ability to expand into a mystical union between Creator God and man.

The infinite combination of within our brain's structural properties enables deep-critical thinking.

Romans: 12:2 states "Be not conformed to this world, but be transformed by the renewing of your {"mind"} that you may prove what is that good and acceptable and perfect will of God". We have a myriad of treasure rooms in our minds!

There are 3 main parts to your brain the {Forebrain], {Midbrain}, and {Hindbrain}. Our Forebrain, itself contains three parts that contain four lobes- {Frontal lobe}, {Parietal lobe}, {Occipital lobe}, and the {Temporal lobe}.

The first lobe located in your forehead {Frontal lobes}. The primary functions of the Frontal lobe are multi-tasking. Think of this area as, the White House in Washington D.C. where the President resides. From this vantage point the brain commands your body via reasoning, problem solving, and judgment and impulse control. This area of your brain also manages emotions of altruism and empathy, and is the organizer-planner of daily activities. We can compare the way this area of the brain functions to the methodology of a Judge, working within the Supreme Court giving legal considerations to all cases that come before him.

Hebrews: 8:10 mirrors this theory perfectly: "For this is the covenant that I will make with the house of Israel after those days, says the Lord I will put my {laws-into-their-minds} and write them in their hearts and I will be to them a God".

The emotion of love also activates the frontal lobe. *1 Corinthians: 13:5 states "Love thinks-no-evil,* which the Greek word "think" means to "reason",

Reasoning and love are connected to the Frontal lobe. *Isaiah: 1:18 states "Come let us "reason" together, says the Lord:*

"Though your sins be as scarlet, they shall be white as snow; though they be red like crimson, they shall be as wool." This is how Creator God's "reasoning" thoughts are expressed in his word, pertaining to Frontal lobes.

The next section we examine contains the {Parietal lobes}. This area processes pain and touch sensation, in the Somatosensory-{the skin and internal organs}. This helps with judging distance and orientation/speech and recognition. The simple act of eating an apple activates the parietal lobe.

Now we go on to the {Temporal lobe} an area that contains the auditory sound sensation. This connects the left brain hemisphere that also contains language recognition. The Temporal lobe also involves the auditory-sound, memory, speech, and smell, also the ability to calculate in a logical manner. Another area that is involved in recognition of spoken words is called the {Wernicke's} with the {Broca's} area which is in the Cortex that monitors speech and facial nerves.

The {Occipital lobe} comes next, since this area controls-visual-sensation and processing-the realm of visual sight, and the area that has the fabric of dreams. Following these various other parts in your brain are the Hippocampus, Hypothalamus, Thalamus, Amygdala. These represent the unique makeup of a world within your brain that navigates through the composition of the midbrain.

On to the Corpus Callosum, which is the neural bridge that connects the two hemispheres of right and left brain together, it is located right in the center of the brain.

The next fascinating area to consider is our {Hindbrain}. This is a part of the brain that connects the Medulla which is part of the super-highway into our spinal cord. The Medulla consists of autonomic functions that our body engages in naturally, such as breathing, digestion and heart function, etc.

The {Pons} is the next area of the brain to consider, that contain thermometer-levels-of-awakening, or consciousness and sleep. Our Cerebellum has the capacity to relay sensory information to and from the brain that impacts learning and coordinating movement. Recently when I watched the Olympic gymnasts display their perfect coordinated balance and movement, I knew they had completely mastered the functions of their Cerebellum.

All of our thinking is done in the Cortex which contains folds or waves in our brain. This maximizes the surface area in a fixed space-very similar to folds in a car radiator-to increase available surface area. Interestingly, if our plump brain were to be unfolded, it would stretch out 2 1/2 feet by 2 1/2 feet - the size of the New York Times newspaper! Research has discovered that charitable giving lights up the brain on an MRI scan indicating giving generates pleasure while simultaneously having effects on altruism!

As I sipped my green tea at a restaurant explaining all this to a computer consultant from India, he bellowed "Are you a brain doctor"? I was honored, but I replied; "No just studying Neuroscience."

CHAPTER 6

BRAIN PARTS AND FUNCTIONS

The area of the brain known as the Corpus Callosum is the largest connector-bridge and houses fibers of over 200-250 million axons. These axons communicate and cross-talk between the left and right hemispheres of the brain. Neurons are a basic but essential building block of our cognitive function, and are integral to the effective working of our nervous system. We have previously discussed the lobes of the brain, now we will look at the components of our forebrain.

The first portion of our limbic system is the Amygdala. This emotional part, encapsulated in an almond shape, assists in storing and classifying emotional memories. We are all aware of Valentine's Day and how the celebrations of that day cause our emotions to skyrocket. The Amygdala distinguishes visual information, facial recognition and complex emotions. It produces fear, and can trigger strong emotional symptoms of sweaty palms, freezing, trembling, increased heartbeat, and increased respiration. In other words, stress hormones are regulated in this area of the brain. Creator God's word states in *1 Peter: 5:7 "Cast all your cares upon Him for he cares for you."*

The Amygdala is also the creative storehouse for our dreams; this unique portion in our brain's treasure trove is full of incredible actions.

The next area that is most intriguing to look at is known as the Hippocampus. This is the Greek word for "Seahorse". Note these two interesting details. 1. This area of the brain is made in the shape of the seahorse. 2. The word "Campus" is used in relation to Universities which, in our society, are considered the seat of all academic intellectualism. The Hippocampus is responsible for knowledge, learning, classification and short and long term memory. It can be likened to the ram or the main drive in a computer. It processes and stores new temporary information for long term storage.

The Hippocampus could be considered the switchboard for the brain. It has the ability to gather individual pieces of information and put them into context. I like to refer to the Hippocampus as the "Pentagon" compiling and storing "top secret" classified information, interpreting incoming nerve signals and spatial relationships for decoding and memory recall. So we need to be on "Orange Alert" and not allow our "Pentagon" to be bombarded or infiltrated by terrorists (i.e. free radical thought processes) that bring us out of aligning with Creator God's truth and pure mindsets! The functioning of this part of the brain brings to mind *Colossians 2:9-10: "That we may increase in the knowledge of Christ who fills all in all things. For in Him dwells all the fullness of the Godhead bodily. And you are complete in Him which is the "head" of all principality and power."*

The next area of the brain to consider is the Hypothalamus. This is linked to the pituitary glands and monitors our body functions such as daily-sleep/wake-cycle. It makes sure our appetite, thirst and emotions run smoothly with balanced autonomic and motor functions. It functions in a feedback loop to enable all of the body's hormones to cycle together. When the Hypothalamus is out of alignment, a hormone is released called CRH. This hormone influences the body's response to all stressors whether physical or emotional. The release of CRH then stimulates the release of an additional hormone called ACTH which releases adrenaline in high pressure situations.

Next we have the Thalamus which is very closely related to the Hypothalamus. Most of our sensory signals such as auditory, sound, visual and somatosensory are released within the internal organs of our body and skin, and they will travel through this organ on their way to other parts of the brain for processing. You could compare this function to Homeland Security Guards at our borders patrolling and keeping out illegal aliens! The Thalamus is responsible for motor control and has an incredible capacity for high definition functions.

Let us take a look at our "stone-age" brain which needs to increase its spatial capacity for learning to stay in step with our "circus-high-flying-juggling-lifestyles"! According to Neuroscientist Dr. Torkel Klingberg in his book "Overflowing Brain", we are being flooded continual torrents of information overload.

Consider the daily demands of a workload that require eight Windows open at once on a computer – releasing an influx of visual distractions, interruptions, and a dizzying amount of continual data piling up like the leaning Tower of Pisa! In California the freeway signs flash "don't-text on cell phones". This information overload limits our working memory and causes an "attention deficit trait."

Technology continues to outpace itself, changing working patterns and causing brain circuits to overload. We are constantly challenged to increase our mental capacities by the impressions that overwhelm us daily…as olives in the oil press! The antidote for these symptoms is clearly found in *Corinthians: 14:33 "For God is not the author of confusion but of peace."* To avoid the head on collision course of disaster that will inevitably result from such a lifestyle, we need to shift into short forays of solitude so our brain will no longer feel like a deep-fried Twinkie-and regroup to smell the roses, thereby resurrecting the gears of our mind into streamlined thoughts.

How do we anchor our over-utilized memories and restore damaged brain cells? We can utilize the wealth of natural and organic supplements at our disposal. One such supplement that is relatively new in the west is called "Lion's Mane", or "Pom Pom." It is derived from a mushroom and because of its long cascading tendrils it actually resembles a cascading lion's mane. Mushrooms are a staple in Chinese medicine. For hundreds of years they have been known as a source of extra

Nutrition on that promotes "nerves of steel and the memory of a lion". We could all use a dose of that, along with a synergy of supplements to supercharge our immune system to destroy those invaders of mental sharpness! Better than Star Wars "The Force Strikes Back"! Lion's Mane is a powerful antioxidant and also an anti-aging food for our brain.

Some of the benefits of Lion's Mane lay in its ability to regenerate brain cells, improve memory function, and stimulate nerve growth factor, while improving mental clarity. Nerve growth factors are a family of proteins that maintain and regenerate human brain cells. Dr. Ward Bond PhD in his TV show "Nutritional Living" states that as we age we stop producing this nerve-growth factor which, in turn, leads to brain disease and loss of mental precision. Interestingly, there was a problem when researchers from Shizouka, Japan had to overcome Nerve Growth Factor in its original state.

They found it could not be used as an orally administered medicine because it did not cross the blood brain barrier (a protective seal over the brain). Later research enabled the team to discover a class of compounds in the Lion's Mane mushroom called Hericenones (sounds a little like Hercules the strong man to me!) These elements cause brain neurons to regroup. This repairs neurological degradation, increases intelligence, and improves reflexes.

There is a passage in *Revelation 5:5 that states "Behold the Lion of the Tribe of Judah."* The royal roar from that majestic Lion's Mane, spiritually speaking, unleashes a penetration into the brain known as seals. Christ name is referred to as the "Lion of Judah." The royal roar from that majestic "Lion's Mane" spiritually speaking, unleashes a penetration into the blood-brain-barrier in the mind's thought processes. Christ's redemptive blood on the cross covers all-as the blood brain barrier covers the brain!

The blood brain barrier, BBB, is very selective in what gets through its almost invincible walls. Researchers in the lab discovered "Chimneric Peptide" which is to help with disease in the brain. Half is a drug which does not cross BBB and half is the molecular "Trojan-Horse" which does. The Trojan horses

are genetically-engineered proteins. In Neuroscience the term Trojan horses are also being constructed to slip through genetic material.By encasing genes in fatty spheres called Liposomes, which are coated with a special polymer which certain antibodies trick the brain-capillary receptors into letting Liposomes pass, where they deliver their "Payload to Brain cells."

I am not referring to Botox on the lips to plump them up either. This reminds me of when I was relentlessly trying to learn to sing Opera and play the piano at the same time. I had such a mental block in my brain that it seemed almost impossible for me to let these musical accompaniments infiltrate my brain's thoughts, at the same time. Then after repeating the mental and physical exercises in each area for years it finally slipped through the mental block that I had for a long time. Today I enjoy the payload of playing and singing opera music and keyboard sounds together before people.

The key to the brain's plasticity is the nerve-growth-factors. These act as "super-nannies" to brain cells. They are present when brain cells are first born, making sure they are nourished, grow, and make the right connections. They are there assisting the newborn life of the cells, guarding their health and repairing damage. They are there to insure that there is healthy formation of brain cells to function in the ways that God created them for, to learn and remember.

As exemplified in *Hosea 6:1 "My people perish for lack of knowledge."* When NGF decline or disappear, individual brain cells start to collapse, shrink with aging. This occurs bcause the NGF is no longer there to protect cells from being executed by the free radicals. Similar to the lamb being left to the wolves, or in our case "molecular-piranhas" fall down on the job, shrink and eventually die. Our tiny soaking sponges in a sea of knowledge begin to dry up and shrink with aging.

The discovery of NGF has opened a door to new generations of therapeutic agents that can revive sputtering damaged brain cells, and generate new cells.

I think of the NGF as a neuron hospital that nurse and assist in birthing new life, or to assist in rescuing dying ones. They can be likened to lifeboats rescuing passengers from sinking ships. Scientists hope to outsmart the blood-brain-barrier as they outwitted the body's immune-defense system, and make organ transplants routinely successful.

Pulitzer-prize winning science writer Ronald Kotulak, wrote in his "Inside The Brain Book." He quotes "Researchers are seeking to establish implantable pumps that bathe the brains center, to halt destruction of Alzheimer's and renew and refresh the memory."

By developing similar strategies Neurosurgeons could easily crossover our brains 'castle moat, of catapults, spies, bridges, and ladders, with piggyback antibodies to sick brain cells providing "docking-ports" for nerve-growth-factors! New research has discovered a startling fact regarding a small number of cells in the tip of an embryo. Its brain cells multiply at an astounding rate and as the new baby is growing in it's mother's womb about 200 billion cells are created in several months.

Their function is to get in touch with the body that is Developing around them and they compete to succeed, in the twentieth week of the baby's life, half of the brain cells die off because they fail to connect to the awakening body. Paul wrote this phenomenon in *Galatians: 4:19 "My little children, of whom I travail in, birth again until Christ be formed in you."*

Retraining your brain to rethink your life causes a release of power to bring about dynamic change. The rediscovering of a hidden archive on how to expand the borders of your mind exists quite simply by uncovering kernels of great new thought processes! The powerhouse cell of a neuron reminds me of a transformational leader, at the helm of a new society. They, the leader and the group, are 'wired' together and 'fire' together.

The Author of "New Breed Leader" Sheila Murray Bethel, PhD states in her writings:

1. Commitment of a leader is to inspire
2. Creation of vision is to fire
3. Creativity, enthusiasm and imagination creates
4. Ideas and actions attracts
5. Energetic committed followers reinforce.

In our neurons circuitry this is a description of how healthy key mind thoughts operate, and how you can step across the tyranny of old mindsets into the freedom of the unlimited horizons of creative new thought processes.

Groundbreaking power thoughts that enable you to pull off impoverished pauper thinking and replace it with rightly positioned perceptions from the Prince of Peace, our Creator God.

BRAIN IMAGE

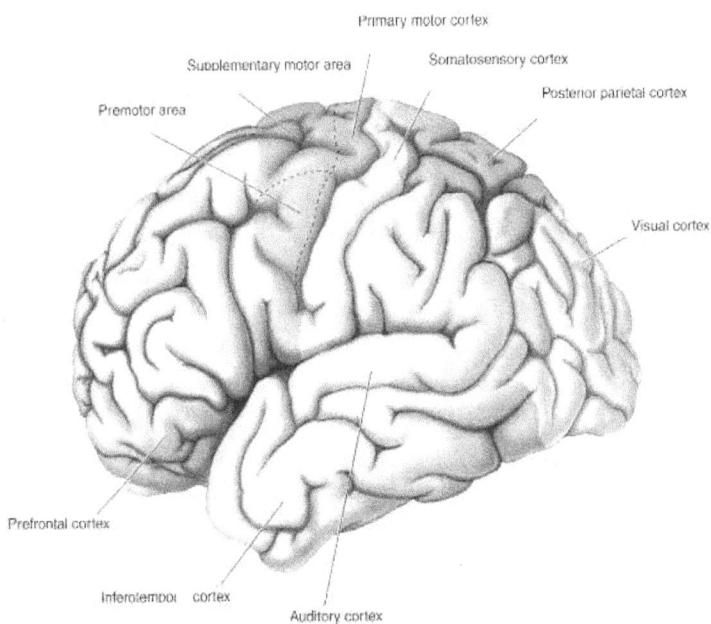

Primary motor cortex

Supplementary motor area

Somatosensory cortex

Posterior parietal cortex

Premotor area

Visual cortex

Prefrontal cortex

Inferotemporal cortex

Auditory cortex

NERVE IMPULSE

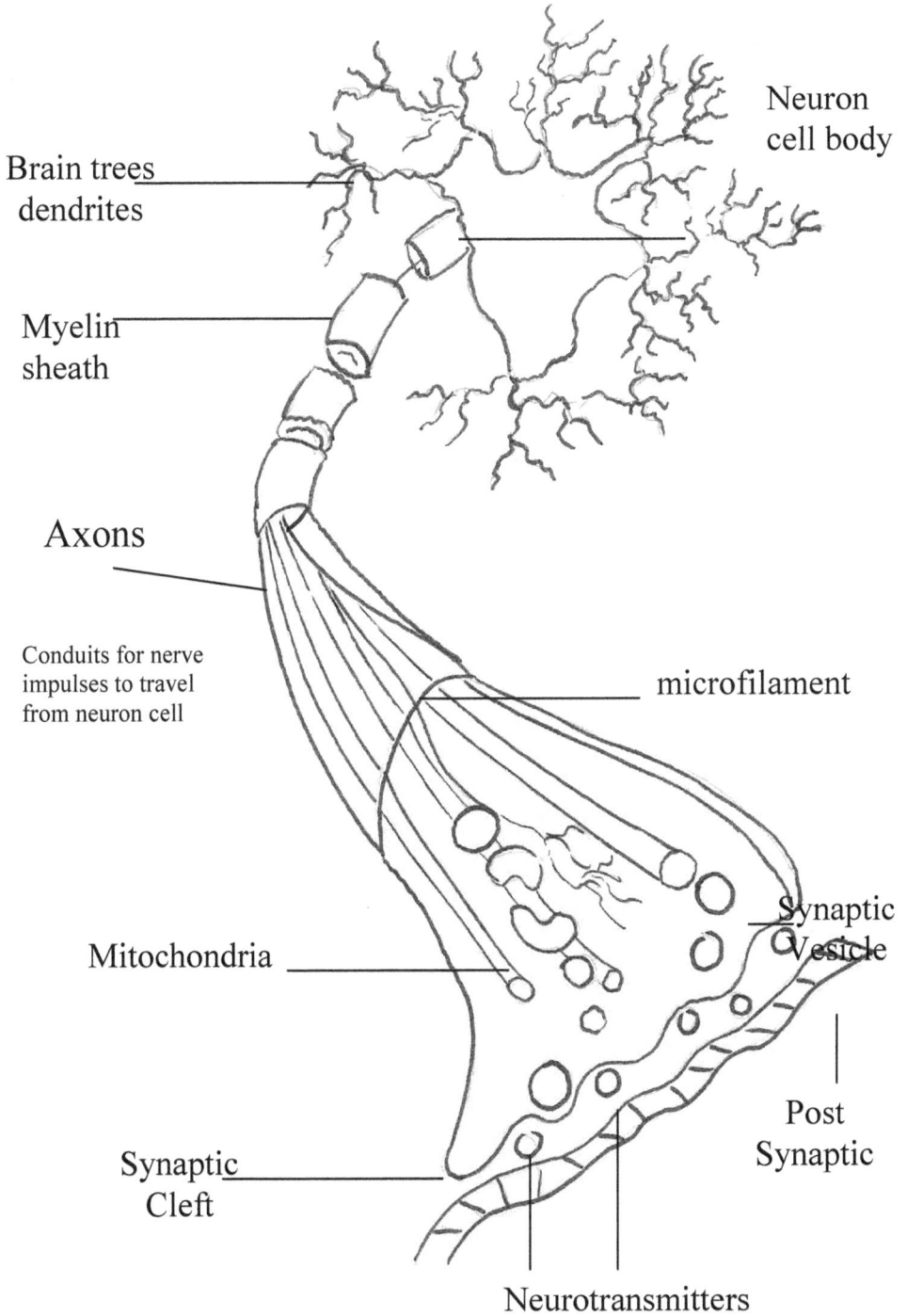

Neuron
cell body

Brain trees
dendrites

Myelin
sheath

Axons

Conduits for nerve
impulses to travel
from neuron cell

microfilament

Mitochondria

Synaptic
Vesicle

Synaptic
Cleft

Post
Synaptic

Neurotransmitters

CROSS-SECTION OF BRAIN

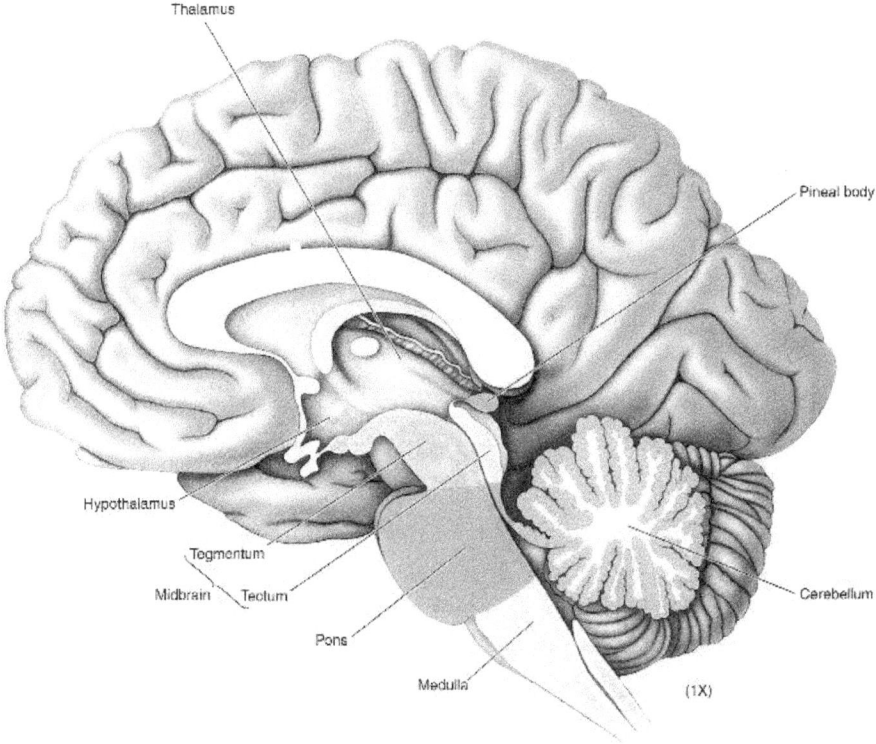

Thalamus

Pineal body

Hypothalamus

Tegmentum

Midbrain   Tectum

Pons

Medulla   (1X)

Cerebellum

CHAPTER 7

BRAIN AUDITORS

Let's take a trip to Vagus not Las Vegas but our "Vagus nerve" which is the largest nerve in your brain. The Vagus nerve is a key component to regulate messages. We can compare this nerve to what a delivery Fed-Ex truck driver does; he simply receives and sends messages. When circumstances arise and we are overwhelmed with too much stress in our brain circuitry, an antidote for this is found written in the word *Isaiah: 26:3 "You will guard him and keep him in perfect peace whose "mind" (both its inclination and character) is stayed on you, because he trusts in you."* Since the Vagus nerve comes to the rescue like superman, acting as an IRS auditor examiner to your body parts.

There are also very important components in your neurons called "Neuro-transmitters." They are located between your cells, for communication, as if dialoging on the cell phone. Amazingly these nerve-transmitters act as conduits to flow in and out of the receptors of the impulse neurons as if they were little walkie-talkies. The right messages with the right keys that fit into these neuron receptors, enables glucose an entry-way into muscles by contracting and giving the nerves a charge! "Eureka-bolts-of-energy"! Sounds like syrupy energy drinks in 7-elevens.

Neuron receptors are what the United Nations are to Global Governments, where everything dialogues information throughout your body. These unique neurotransmitters are facilitated with different functions, and convey various moods and feelings such as excitement, relaxation, memories. How we feed our brain with proper brain diet to keep it fueled properly affects our production of neurotransmitters. Since nutrient supplementation can correct and enhance mind, mood, and memory behavior. All major neurotransmitters are made from amino-acids and dietary-proteins. One of the mistakes of a low protein diet is not ingesting enough amino acids to make enough brain-Neuro-transmitters, with a cause and effect of an alertness decline. The remedy to alter this is a "high protein" diet increasing your alertness. Amazingly we have over 50 neurotransmitters and we'll look at four of the major ones, because they are the essentials of life for your brain.

In Dharma Khalsa M.D.Book on Brain-Longevity, he has researched the facts that expounds on the first neurotransmitter which is called "Acetylcholine." This assists as conduit of memory and thought. If you don't have enough of this you'll experience memory loss, probably the most extensive cause of age-related cognitive impairment is Alzheimer's. Lecithin is a good source for this, which is found in eggs, also Choline nutrients, is great and plentifully supplied in multiple vitamins which will help you considerably. Another supplement is called Huperzine A, which supports learning and memory for boosting the brain and will help to bring restoration by helping to prevent neuronal memory loss in it. The 2nd neurotransmitter is "Norepinephrine."

This releases a burst-of-energy which is similar to adrenaline; this conducts the nuts and bolts for long-term memories, which shift into the storage of your Hippocampus and Amygdala. This is one of your brain's "treasury-of-happiness" chemicals, lifting your moods and energy, without this, you'll experience lower states of moods, a blah feeling, may occur.

The third neurotransmitter is "Dopamine." This brain chemical controls body movement, moods, burns fat, and enhances immunity. The Dopamine and Norepinephrine are structured amino-acids and building-blocks loaded with folic-acid, magnesium, vitamin C, and B12. They are found in high protein foods such as soy, poultry, seafood, and dairy products. If there is a lack of these neurotransmitters you will feel depressed, and cannot concentrate and focus on allocating memories.

The fourth neurotransmitter is called "Serotonin". This makes you "feel-good." This monitors learning, memory, mood behavior, and helps you sleep. If there isn't enough you'll tend to feel blue or emotionally down. It is a nutrient loaded with supplies from the group of amino acids called Tryptophan. That's why when you're feasting at the family Thanksgiving; your turkey that is packed with tasty relaxing tryptophan lured you to sleep. This food group mix of carbohydrates satisfies your taste of feeling good, and is loaded with brain chemicals, for increased performance. When you feed all four neurotransmitters this will enhance and help control mood swings.

To keep you "sizzling sharp" all day, by keeping fuel in your brain! One thing that can really damage your brain-trees-dendrites, and rob you of comprehensive skills is continual bouts of low-blood-sugar! Number 1 enemy of the brain which runs on glucose. Surprisingly when this drops, within minutes your brain's power plant begins a spiraling-disconnection of brain trees, or so-called a power outage, and fragmented thinking or mental fogginess settles in. This will damage neurons. Most of us have lost millions of neurons just from low blood sugar which has a jarring effect on memory traces to your brain?

This causes a domino effect, when you miss meals. You might feel lightheaded and have poor long and short term memory recall. When you maintain a high blood sugar flow, in the diet this will enhance an active mind with a surplus of healthy neurons. Also when I meditate on Creator God's word, my mind is stretched to increase in *Psalm 34:8 "Oh taste and see that the Lord is good; blessed is the man that trusts in him."* When we taste of the Creator's spirit we are renewed in our thought processes as well.

You might wonder how Neuroscientists could find a seat of awareness in the cacophony of billions of jabbering neurons. It is the unbridgeable gulf in our entire neurological system that is protected by a "Blood-Brain-Barrier" which is tightly packed cells that act as a filter system. Their appearance is a shape of a picket fence; a border shield that encases and surrounds the brain, as a protective helmet.

I think of the dominion of Christ's supreme sacrifice in *Colossians 1:20-21 "And having made peace through the blood of his cross, by him to reconcile all things to himself: by him whether they be things in earth or things in heaven. And you, who were sometimes alienated and enemies in your "mind" by wicked works, yet now he has reconciled you."* Christ a spiritual type of the "Blood-Brain-Barrier" with practical application of the sovereign word will help to change your thought consciousness in your mind, by cleansing your thoughts as if through a filtering system.

I stayed at my sister Barbara's house and weekly would venture to the grocery store and filled her 5 gallon water jugs for home use. Advertisements plastered all over the vending water machines displayed "filters checked daily in this reverse osmosis process would facilitate machines to have clean filtered water.

This is what our blood-brain-barrier does in protecting the brain and its functions. As an osmosis of filtered-flowing-water can purify streams you can also wash away thoughts by the cleansing scriptures with the word, described in the passage *Titus 3:5 "Not by works of righteousness which we have done, but according to his mercy he saved us, by the "washing-of-regeneration, and renewing of the Holy Spirit."*

An act of grace, even when Christ washed the disciple's feet, for he said unless I wash you; you have no part of me and in *John: 13:9 Simon Peter said unto him, Lord not my feet only, but also my hands and my "head."* He had an understanding that he knew he had to have his mind washed and cleansed in his thought patterns.

Even so the blood of Christ covers and washes us in our thoughts to keep us on God's pathway, inadvertently as a coating of the {blood-brain-barrier} in brain parts is sealed as a protective guard. For example the Royal family resides in London at the Buckingham Palace. The Royal guards are stationed as a protective hedge to guard the "Head" Monarch Queen Elizabeth, arresting any violators of Royal law that seek to usurp her noble authority from the throne.

This is portrayed in Genesis when Adam and Eve fell from their fellowship with God by the deception of Satan revealed in *Genesis: 3:15 "And I will put enmity between thee and the woman, and between thy seed and her seed; it shall bruise thy "head" and you shall bruise his heel."*

Which is what Christ did upon the cross for us in *John: 19:17 "And bearing his cross went forth into a place of a skull which is Golgotha."* When he said ("It Is Finished") he broke the "Headship" of Satan's domain forever.

CHAPTER 8

THE BRAINS MULTIPLE MAPPING

According to the Drake Institute there is a process known as Brain Mapping. They place 19 sensors over the surface of your head, which records the brain wave activity. This reminds me of the huge pipeline waves on the island of Oahu in Hawaii's north shore where surfers ride the Big-Kahuna-Waves. Let me explain. We contain many variable wave-lengths not on the island, but in our mind and mapping it will record electrical-activity, that will not affect your brain. It simply forms a data base that shows how an individual's brain should function at any age. The Multiple-Mapping views the 19 different locations to determine whether the brain waves are to slow or to fast.

For example if a frontal region of the brain shows excessive slow brain waves and the individual has a problem with attention deficit, then the diagnosis might be one or several neurotransmitter deficiencies. In this case an adult could use "brain boost" vitamin supplements to solve the problem. For example {GABA} is an anti-anxiety amino acid and used with other amino acids are excellent for symptoms of ADD in children. This blocks the bombarding of firings of excitatory messages, so the brain doesn't become overwhelmed. GABA, glutamine amino acids, and glycine are proven for vital energy and smooth running of brain functions. The glutamine has over 21 amino acids, which I take daily as an alternate fuel source for the brain when blood sugar is low. It is located in the Hippocampus, which is the seat of all learning, intelligence and memory.

According to Dharma Khala M.D. in his study of Multiple-Mapping in the brain, an individual's thoughts about his personal possessions such as a sapphire blue car in their garage might be located in different places in the brain. These assemblies of small network neurons cradle thoughts about cars, in another area of your brain that cradles blue colored things, and yet another place remembers things in your garage.

I saw a television show on Jay Leno where he displayed his collection of prized antique cars, stored in a vast warehouse. He really had to have multiple mapping to keep track of his unique cars. The dynamics of multiple mapping is the fact that memories must travel in a circle, in our "brains transportation system" with the neurotransmitters. It is relatively hard to kill a memory, because they exist in many widely separated areas.

An example of this is when I moved from Texas to California I put all my personal things in a U-Haul unit until I would need them again. In retrospect your brain's Hippocampus is compared to a U-Haul storage unit that can ship memories in long term storage.

Years ago in the famous Indiana Jones movie, "Raiders of the Lost Ark" had a scene at the end of the movie, where they moved the lost "Ark of the Covenant" and hid it in a top-secret-storage to study its future mysteries.

Donald Hebb proclaimed that memory stems from neurons working in unison to strengthen the synapse where they meet. We're sociable beings, even on a cellular level where active neurons cement their mutual bonds, forming "little-cliques, within our brain." Neurosurgeon Frank Vertosick explains. "They really are social clubs, these tiny societies of cells. Some will be influenced more than others by lobbying neurons. Just like our human society, the majority reaches a decision, despite naysayers."

Altruistic neurons act foolishly outside of the clique, they act separately, selfishly, to promote their own genes, oblivious to the others, and it doesn't give a whit if the others aren't on the roller-coaster, and that's no brainer!

The brain is organized into 40 physical maps which govern vision, language, muscle movement, and hearing. How maps are organized is influenced by electrochemical signals flowing through the brain, in adolescent early years. Your brain-gates are a neural interface with the Enigma-Eternal-Genius, the everlasting Creator God. That is identified in the scripture verse *Matthew 22: 37 "Love the Lord God with all your heart, and with all your soul, and with all your mind."*

As you release these thoughts over your life this will Enable you to think in a freedom route similar to a GPS mapping system in your car to help you get to your destiny. Our thoughts wind through a maze of open windows in your mind, into a line of demarcation thoughts, which is centered in *1 Corinthians 2:16"For who has known the "mind" of the Lord that he may instruct him? But we have the "mind" of Christ."*

Your Cortex brain is made up of 6 layers of folded gray and white matter of cortical columns, an abundant activity of inter-neurons in a "Cross-Talk" between one another. They also visualize each other as a "cross-modal" set of neural firing that gives you a crowning-cortical-capacity. Daniel Siegel's book called the "Mindful Brain" describes your thought signals in your left hemisphere brain as traveling across the Corpus Callosum, which is your neural bridge, into the right hemisphere and vice a-versa, to release the language of cross talking to each other.

A visual picture of this is referenced in the earth's center known as the equator which separates the northern and southern hemispheres. This is a metaphor, of the right, and left brain hemisphere, crossing over into its visual field. Therefore it's perception into the left hemisphere will cross-talk back and forth. Similar to an axle under your car that connects your right and left tires. If neural damage is done by severing the rope-fibers in the Corpus Callosum neural bridge, it will cause a crossover in 2 halves of the brain's right and left hemisphere's, which is diagnosed as "Split-Brain."

Unfortunately this causes two separate minds to operate, and sphere's of consciousness. Creating a mental division in learning memory, and a reduced ability to name, and identify non-linguistic environmental sounds, even though the right brain is willing to assist the left brain in its activities.

Even when plagued by conflict because of severing between the two hemispheres, there isn't enough exchange of information. This is defined in *James 1:8 "A Double-minded man is unstable in all his ways."* Since this will cause an exploding war between the two sides to unfold. The remedy for this is in the word *Ephesians 4:23-24 "To be renewed in the spirit of your mind and put on the new man which after God is created in righteousness.*

CHAPTER 9

⟨ᑕᓄᕇᔱ⟩

HIPPOCAMPUS
REFLECTIONS OF THE SEAHORSE

The Hippocampus part of our brain is shaped like a seahorse and in the Greek language, hippo actually means horse. This tiny area is where our short immediate and long term memories are stored. Remember I called it the Pentagon in an earlier chapter, the intelligent seat of learning.

My adventures in California brought me flashbacks of my visit to the Long Beach aquarium to see the rainbow-splashed-seahorses floating in their oasis of saltwater. Daydreaming in my mind I wish I could bridle the seahorses and ride off into the pacific sunset.

This reminded me of a walloping rodeo year in the Wild West Texas. I settled in a small town named Denton Texas and lived by a ranch full of show horses. Not too far from former President George Bush's ranch. So I loaded my digital camera with a portfolio of magnificent photographs that I shot of the horses.

They trotted up to me as big puppies wanting to play, so I reclined on my back and took pictures of them from the ground up, where they whinnied with wet-noses that would sniff and nuzzle me on the meadow. Probably in search of carrot treats, as I zoomed my camera at a grass-roots-level and shot their hooves before they could stomp on me. My ground photography session of these majestic stallions will forever be lodged in my Texas memory-banks, thanks to this part of my brain called the Hippocampus, a seahorse in my tank of fluidity intelligence. The codex of my brain atlas has filed countless stories of gold panned memories. One memory in particular was my travels through Europe particularly Munich Germany during their harvest time called Oktoberfest.

Jolly lederhosen clad fat man waddled around the festive streets with his massive Clydesdale horses. He bellowed out to me "Fraulein" come over here and I'll take your picture"! Jovially laughing as he handed sugar cubes to his horses that they gently gummed off with their enlarged-lips that tickled. An image forever embedded in my memory was his horse that you could tell was accustomed to his laugh.

I posed, and "flash" he snapped my picture. Suddenly his snorting horse turned and licked my arm "Eek-Gad!" I moaned as the horse's gurgling fountain of slobber dribbled over my arm! Suddenly there was a sound of siren-pitched-noises that echoed through the costumed lederhosen crowds. With a chorus of baby horses that whinnied an aria coloratura, it was a celebrated memory vaulted deep in my conscious mind.

I had the unusual adventure of relocating from suburban Orlando Florida to a small farm in Texas was great for growing organic vegetables. Within a stones throw of walking distance, was an abandoned horse pasture with a hidden orchard. When it was winter it became quite beautiful with snow quilted trees in a winter wonderland. This reminded me of the tree lighting ceremony in Rockefeller center in New York! The orchard would take on new beauty with the myriad of changing seasons, One late summer I moseyed over to the orchard where weeds had grown and hidden the rustic wobbly gated fence.

I always walked with a stick for pest control. Suddenly to my surprise a hidden "Texas Rattler" shot gunned upwards towards me lunging towards me like a lightning bolt. I ran towards the menancing beast shrieking at him! A rumble quaked through my thoughts. "Remember the Alamo"! Take your stand against the villain. With his Hissing life-threatening strike, I swiftly whacked his reptile head and he slithered away leaving me the precious fruit, dangling in high tree branches of scented peaches, and spiked granny apples. This reminded me of the scripture *Genesis: 2:9 "And out of the ground made the Lord God to grow every tree that is pleasant to the sight, and good for food; the tree of knowledge of good and evil."* As we know the serpent coerced Eve to eat the forbidden fruit, thus the eyes of Adam and Eve and their understanding were enlightened to the spirit world which was hidden behind a veil in their mind. This opened their minds up to the knowledge of good and evil that was an invisible world to them before this time.

For every predicament there always seems to be antidote and fortunately God provided the most crucial remedy for mankind in time in *Genesis 3:15 "And I will put an enmity between thee and the woman, and between thy seed and her seed; it shall bruise thy "head" and you shall bruise his heel."* This reminded me of my adventure in the orchard garden, where I bruised the "head" of the serpent. That would keep me from my tree of life, where I retrieved baskets full of peaches and apples that I toted home for my delicious pies.

Recently I was in California and rollerblading through a park area, and spotted a flock of wild lime green screeching parrots that had flown into a massive apricot fruit tree. They began to devour the tiny fruits as I cruised under the branches observing their shockingly bright colors. Suddenly the flock started to bite off the apricot branches and all the fruit started bombarding me in the head. I swiftly grabbed up the precious bounty, thinking I was rained upon with fresh fruit on my mind, what an awesome experience as I carted them back home.

Various types of memories live in different areas of the brain, and circled groups of neurons added together, form a single memory. A good way to picture this is described by Neurologist Jeff Victoroff, who implies a football game where 10,000 people are sitting across from you, each holding a colored card. At a signal, they flip their cards into position, and the pattern reads "Go Cowboys!" The idea "Go Cowboys" is not written on any one of those 10,000 cards, neither is it located in any one seat.

It's created through a pattern of activity, as coordinated firing of 10,000 neuronal responses. Likewise our memories are stored in our brains not in any one particular place but as a distributed network of neurons, cued to turn their cards of synaptic activity in a unified way. These firing neurons work with unprecedented accuracy in a vault of millions of memories on replay like your MP3 player.

My Native American friend Mando makes the most aromatic scented oils for every occasion, a potpourri of exclusive fragrances for healing. As a Native American drummer, he will pour these scented healing oils on his drum. So while he is playing the drum, the entire room is satiated and permeated with these healing scents. In a large drum circle of various tribes, I was intoxicated with aromatic fragrances that wafted through the booming room. Growing as sounds and scents resonated through my mind. Thundering more profusely, a purer intensity waved through the crescendo of scented drums.

A river of textures, atmospheres, sights and sounds shattered through them. When you give gifts of perfume, flowers, scented oils, or soaps you are really giving them a gift-wrapped-memory! Smell adheres to memory, and is connected with many emotions. After all we have over 10,000 scents of smell! Most sensory functions proceed through our Thalamus first; a gateway to the Forebrain, but when you sleep the gate closes. Though when you are asleep your nose will detect a scent and it will career past that route and send a message on to the limbic system, a unique part of our brain. Both the Amygdala emotional memories and the unique makeup of the intelligent seahorse Hippocampus that corresponds with memory-and-knowledge become the CNN news reporters! As Solomon the King penned in *Proverbs 10:7 "The memory of the just is blessed."* When we inhale different fragrances from various scents such as perfume, flowers, animal, mineral, ocean, or air, our mind has a cause-and-effect to awaken so many memories, because fragrance evokes strong images and emotions.

In some airports in Asia they have a scent of freshly mown grass that wafts through the hallways producing a calming effect on your mind when you arrive from your hurried plane flight. Meanwhile various scents veil us in rapture, because we each have our own aromatic memories, in part because smell awakens learning.

Rachel Hertz who studies smelling at chemical senses center in Philadelphia, did an experiment with volunteers where they inhaled certain smells while looking at art paintings. A few days later into the experiment she gave them the same smell, and they responded with great clarity, remembering the exact painting, associated with it. So their mind was stimulated and the memory was awakened, to beautiful art by associating and linking fragrances to it.

CHAPTER 10

GRAY MATTTER IN THE BRAIN

The extraordinary efforts of our mind's government can surmount and recover the brain. This armory of intelligence made up of gray matter in our brain is nerve-cells and white matter is long spindly filaments extending from cell bodies that resemble electrical wires transmitting signals. There are portions of Gray matter in specific regions of our brain, which is highly heritable, having a strong environmental influence. Increased Gray matter reflected at least in part of professional musicians, showed the number of years devoted to musical training in the Broca's area.

In our aging process after 45 years the gray matter in the brain seems to decline and shrivel faster than a prune to a raisin. The good news is because of the brain's plasticity, you can change when you embark on becoming a lifelong learner, you will decrease memory related problems. This will help to increase your serenity of mind, and clearness. One thing that helps me is my favorite elixir of green tea mixed with Ginkgo Biloba, which is loaded with antioxidants.

Christopher Columbus was the ultimate immigrant pioneer, who embarked into aquamarine waters only to discover a new land as he expanded his horizons from a localized way of thinking into a new utopian environment. When I relocated across America, navigating through a maze of uncharted territories, from California to Florida I was surprised with panoramic vistas of new veritably diverse lands. Traveling through each state from the Grand Canyon, New Mexico, Texas, and New Orleans, Alabama, Georgia and beyond was an epiphany for me a true awakening experience. The epithet of how great our nation really is! A myriad ocean filled with epoch golden cultures, people groups, tribes, nationalities, from state-to-state. So are our diverse brain-states. We can experience a localized way of thinking, nationally, or globally.

This is amplified in *Colossians 1:18 "He has been made the "head" of the body the church who is the beginning of the firstborn from the dead that in all things he might have preeminence."* He is the ultimate supreme Head Ambassador in his infinite universe, which he has made us to be one with him. This reminds me of the powerful Gamma Rays that pierce these earthen vessels of our mindset. Gamma Rays are the highest frequency of light energy levels in the universe that are "pure electromagnetic energy". These wavelengths which travel beyond the velocity of the speed of light. A Discover Magazine reporter wrote an article that detailed the report from researchers that uncovered "Magnetite" in sampled areas of the brain.

"Magnetite Crystals are present in almost all living organisms. Scientists found that in different parts of the brain were "Magnetite Crystals" that were consistent in size, shape and distribution, suggesting they have biological functions. No one knows where the {tiny-magnets} are located in intact cells. One possibility is that the Magnetite Crystals could be coupled to Ion Channels that regulate the flow of materials in and out of cells.

When exposed to strong electric fields, the {tiny-magnets} could reorient themselves and either open channels or slam them shut. In a similar way the Panama Canal locks, overflow to the brim with water, allowing ships to traffic through the narrow canal and simultaneously rise from sea level into a higher height rising chamber which allows passage through narrow channel canals.

The mineral Magnetite has Iron Oxide in it which plays a big part in the field of magnetism which carries electric-vector-electricity which produces magnetic fields. That would affect your brain cell's healthy activity. In nature this is typical of fireflies or lightning bugs in the Midwest with their flashing vectors of electricity that light up at night. All of which bring us back to the magnetic love of God, like a compass pointing us in the right direction. Another interesting study was completed in the London Taxi School that involved taxi cab drivers which had to learn to navigate the city's road maps.

After research there was an increase Gray Matter in the portion of the brain called the Anterior-Hippocampus, known as the "seat of intelligence" and learning. This enabled taxi school students to increase in their spatial navigation which helped to accelerate their driving experience. Older adults that were involved in aerobic training multiplied new brain cells and also increased their Gray Matter compared to those who didn't do cardio-exercise. Huffing and puffing with sweat dribbling into my pink headband; after 20 years of absence from aerobics, I began the new aerobic class with an instructor a rippling rendition of Whoopi-Goldberg. A drill marine-sergeant barking orders, to get my-noodle-muscles in shape in beginner "boot-camp"! I felt my brain-trees increasing as I crawled out of the class! Our magnificent brain at rest contains a host called "Baseline Activity." Within our awakened state we use 20% oxygen even though it makes up only 2% of our body's mass.

Baseline activity has many features with its marvelous power to restructure knowledge in the mind's background which will simulate future states and events thus manipulating memories. For example Kelly surmises in her mind by a dialogue she rehearses in her daily planner of her deep thoughts within herself as she is contemplating her anniversary for the following week to go hot-air ballooning. Kelly filed that thought into her baseline-activity of background memory as she daydreamed about flying away upwards in the balloon

Information pulses thru the eye-gates because they each contain 125 million visual receptors in each eye and compete for the brain's attention! With this reasoning and planning our emotions and drives can occur with no external stimulus that has entered our eye-gates. In **Genesis 1:26 "God said let us make man in our "image" after our likeness."** The first known "Neuroimaging" Creator God did was in paradise after making his likeness in Adam and Eve".

Neuroimaging is envisioning the act before it happens. This show of activity decreases in the same brain area, just before a person performs a goal-directed task, for example the Olympic gymnast before their feat of gymnastics, the boxer in his dressing room before entering his prized-ring-fight, an actor before his stage production, the singer prior to performance, all encompass every curlicue of the Cortex brain.

CHAPTER 11

❧✦❧

BRAIN WAVES

We have a phenomenal amount of interconnected nerve cells which amount to over ten billion that operate at the same time. If you had enough scalps hooked up you wouldn't get a "lightning bolt" but you might be able to illuminate a light bulb "Eureka"! Jesus said in *John 8:12 "I am the light of the world he that walks after me shall not walk in darkness but shall have the light of life."* This living light bulb illuminates with incredible love, light, and healing.

We have pockets of energy that sweep through our brain waves in four different patterns. The first pattern is the "Beta-wave" which is engaged in strong mental activities, ranging from 15 to 40 cycles per second. Gamma waves are grouped with Beta waves. These two brain waves contain much visual acuity, with analytical problem solving, judgment, making decisions within our world around us. The Gamma waves are associated with alert, busy, and anxious thinking, fueled by active concentration. The different examples of this is working on the job, Presidential debates, students taking a test, Larry King talk show as he engages with his guests in dialogue's, increases the Beta-Gamma rhythm. The idea is that Beta rhythms might be to link different brain areas together.

Whereas high Gamma waves within a region bind groups of cells together, in timing-rhythm where their information can be sent out, or received. Reminds me of traffic air control towers that allow planes to come and go on a runway landing strip. The next wave is the "Alpha wave" which is in our brain operating in 9-14 cycles per second. This pertains to relaxing, reflecting in a slower but also higher state of mind. In this wavelength people feel more a little more at ease and calmer. This is generated when you are reclining, sitting down, or reflecting with prayer, meditating, going on extended garden walks, or short breaks from work, and busy schedules. It appears to bridge our conscious with the subconscious. Jesus said I am the "Alpha and the Omega" which means the beginning and the end.

Our third wave is {"Theta waves"} which slow down to 5-8 cycles per second. The Theta waves consist of daydreaming. When you are on an extended trip driving along a long stretch of a freeway, and you can't recall your last 10 miles that you just wheeled down the super highway, you were daydreaming in this brain frequency. In this wavelength you are naturally prone to have creative plans and have ideas surface. Some individuals break out in their morning showers singing, others find a euphoric place of prayer. This is a gate-way to mental relaxation, idealization and leads to deeper states of consciousness, which is revealed in scripture of ***Psalm 92:5 "O Lord, how great are your works! And your "thoughts" are very deep."***

This is God's revelatory thinking; regarding inventive good imaginations, which is described in *Psalm 94:19 "In the multitude of my "thoughts" within me your comforts {compassion} delights my soul."*

The 4th wave is called "Delta wave." It is a place before the mind's eye, where we become receptive to information beyond our normal conscious awareness. We increase Delta waves, in order to decrease our awareness of the physical world. Another way we can probe into the drowsy Delta wave is to imagine your driving a car and you shift into first gear. You're not going to get anywhere very fast, so this represents a Delta state. This tends to be our highest in amplitude but slowest in wavelengths. It is a gateway into empathy, learning and memory, and it rejuvenates our brain when it is fatigued.

You might feel as though you're flying in this brain wave. It has uncanny ability to integrate and let go, with reflections of our unconscious mind. In a dream state, though your mind might be in a deep dreamless state. So when you do dream, it is in 90 minute cycles. This is described in *Job 33:14-15 "God speaks once yes twice, yet man does not perceive it. In a dream or a vision of the night when deep sleep falls upon men slumbering in bed. Then he opens the ears of men and women and seals their instructions, to hide man from pride."*

In Japan a group of teamed scientists at computational Neuroscience laboratories have created a device that enables imaging, where thoughts and dreams experienced in the brain, appear on a computer screen. This discovery paves the way to unlock people's dreams. It is the first time possible to visualize what people see directly from brain activity. With this technology, there is a possibility to record and replay subjective images that people perceive like dreams.

In our eyes retina we obtain image recognition which converts into electrical signals sent into the brain's visual Cortex. There was a popular TV Show on a few years ago with the star actor Lee Majors who played the "Bionic Man." His eye retinas would reveal information about him. Interestingly there is a word about Creator God being able to read the unconscious mind in *Daniel 2:28-29 "But there is a God in heaven that reveals secrets, and makes known to the king Nebuchadnezzar what shall be in the latter days. Your dream, and the visions of your head upon your bed, are these; O king the "thoughts" came into your "mind" upon your bed, what should come to pass here after: and he that reveals secrets makes known to you what shall come to pass."* Creator God unravels mystery-dreams in an explicit way!

Our living experience is a choreographed exquisite dance played by the Alpha-Beta-Delta-Theta brain-waves. As William James quoted "I am for those tiny invisible loving human forces that work from individuals, creeping through the cranium world like so many rootlets or capillaries". Trusting your brain can't be earned in a day but trust in Creator God, causes growing faith to quiet our brains doubts. Enabling the mind to reveal and utilize its powers to the greatest extent, which helps develop willpower for a goal to see you through. This is revealed in the passage *Proverbs 3:5 "Trust in the Lord with all your heart and lean not on your own understanding."* Then you will extend your brain waves states into a less stressful life.

CHAPTER 12

BRAIN FREQUENCY: MUSIC, LIGHT AND SOUNDS

Alfred A.Tomatis M.D. an ear, nose, and throat specialist born on Christmas day 1920 and passed away on Christmas day 2001. He pioneered many books, but one in particular called "The Ear and Voice" uncovered the specialty of harmonics in our voice that only the ear is able to perceive. Harmonics charge and energize our brain's functioning state of alertness and attention. This causes us to survey our surroundings with thoughts of physical and verbal transactions in our environment.

Dr. Tomatis wrote that the key for listening is to perceive all frequency sounds in our ears to hear the auditory spectrum. This will enable our right ear to connect to our left brain where language is processed in a fast and accurate connection. The left ear is connected to the right brain, where language cannot be processed. All the neurons have to jump across as miniature-bungee-jumpers via the connector-bridge between both sides of our brain, called the Corpus Callosum.

Based on thousands of tests he had performed done on his clients, Dr. Tomatis reasoned to improve someone's listening curve. Such as the world famous Opera singer Maria Callas

By improving her voice to adapt to different sounds and frequencies she increased her performance as an acclaimed Opera singer. He also treated the renowned actor Gerard Depardieu using the technique called "Gated-Music." This is used with people that have an ideal listening curve which enables them to learn easier. This is linked in the scripture *Psalm 100: 4 "Enter into his gates with thanksgiving and into his courts with praise."* This reference is also found in *Psalm 24:7 "Lift up your heads, Oh you gates! And be lifted up you everlasting doors! And the king of glory shall come in"!*

Doctor Tomatis found that higher frequencies transmit more auditory sound information to our brain, thereby stimulating it more. Some of these frequencies can be found in high pitched arias accompanied by shrilling opera voices. His discoveries revealed certain bands, or zones located within sound frequencies.

A category-zone-three is known as a spectrum of higher-frequency-sounds. Once these sound frequencies were released there was an explosion of energy, intuition, ideas, spirituality, creativity, and auditory cohesion. When our brain hears high pitched sounds, a stimulation occurs which will result in more energy. Since new sounds break old cycles our ears act as "Gatekeepers" and are explosive-dynamos to the brain!

Another side-effect of category-zone-three is that motor skills improve. Good learners have ears that capture energizing frequencies of that spectrum. Poor learners do not benefit from sound energy. You can actually strengthen your brain with sound waves working with light waves.

You can synchronize your right and left brains, and unlock creative breakthroughs. This foresight can shift your mind into Quantum Leaps. For example when the thundering sounds from God they are described as both light and sound. *Psalm 29:3 "The voice of the Lord is over the waters the God of glory thunders. The Lord is over many waters the voice of the Lord is powerful; the voice of the Lord is full of majesty."* How awesome is the frequency sounds of the voice of the Lord as exemplified in scripture.

CHAPTER 13

BRAIN AND MUSIC

Picnic table cloths checkered the Hollywood bowl lawns under a tented galaxy of stars, on a mid summer night. I opened the windows to the mansions of my mind, and heard the albatross of soaring sounds enabling me to hear a cadence of orchestral melodies. While ethereal music echoed down the hallways of my ears my Temporal lobe became hyper-charged! Neurons knuckled each other as "lightning-bolts" activated the crescendo symphony. For a moment, visually imagine your brain as an orchestra dividing into different creative sections.

My brother-in-law plays Native American drums, creating indigenous-sounds of "First Nations" pulsing rhythms. Each musical instrument in an orchestra has a very specific and important independent function; yet it must work in harmony with the other sections to create the perfection of sound, within the symphony, minus any discord. When playing music, each section waits for the conductor to indicate when to start, stop, release, etc. When he raises and moves his baton in instruction all are in harmonious agreement. If a drum section or a strings section has not been practicing, they don't harmonize with the orchestra.

The overall sound of the music seems discordant or out of tune at certain times. This is a better model of how our brain works, as a tightly knit "neuron-orchestra". We use to think of the brain as a huge computer; but it is really like millions of computers working together. The Brain Symposium held a Neurological study of music, with an orchestra and Opera singers in Salzburg Austria. The finest skilled musicians volunteered for the draft, for unique MRI-brain-scans. It was discovered that they had a higher concentration of Gray Matter {memory cells} located in the Temporal auditory area of the brain. This brings to mind the verse in *Colossians 3:2 "Set your mind on things above not on things below."* In reflecting on our "shimmering-halo" of brain-waves, we see that the Alpha brain waves meet Alpha sound waves. Jesus said "I am the Alpha and Omega the Beginning and the End".

A special research project called The Brain Special Project was conducted at Harvard. A window of revelation was opened when researchers exposed a metrical grid of "tempo/phrases" in songs. The grid formed natural "tree branches" which are called "Brain-tree-branches." They were similar to a microscopic forest of trees with no leaves on them. These transitional elements allow speakers and musicians to advance from one linguistic, or musical scaffolding to another. Like Michelangelo when he ascended onto upper scaffolds to paint the ceiling of the Sistine Chapel with spectacular new visual creations. An example of this is when two tuning forks of the same frequency are in a close proximity, they will produce the same sound if only one is struck.

In the same way when our brain hears "new sounds" it generates new frequencies. Science has discovered that when new sound frequency waves enter into tiny cells they have microscopic gates that open up. Other parts of the brain then resonate into a balanced frequency, thus enabling neurons to cut through previously unchartered pathways of cognitive function.

When King David composed *Psalm 27:6 "Now shall my head be lifted up above mine enemies round about: therefore will I offer in his tabernacle sacrifices of joy. And I will sing yea I will sing. Yes, I will sing praises unto the Lord."* King David's mind (or brain thoughts) was elevated as he sang praises to the Lord. A correlation to that scripture can be found in *Psalm 16:7 "I will bless the Lord who has given me counsel my "mind" also instructs me in the night season."* After he blessed and praised God, counsel came to King David and released joy, revelation and inspiration.

NASA Space Center has been able to record "space-sounds" from various planets such as Mars, Venus, Uranus, Jupiter, Saturn's rings, the Sun, Pulsars, and Black holes. Florida's Neuroscience Brain Institute found a larger puzzle in the higher brain region. They uncovered that sounds evolve into specialized coding-strategies in the auditory system {Temporal-lobe) of your brain. These sounds are interpreted and understood in a population-of-cells unlike any other in the nervous system. These special cells can detect sounds in a 360 degree space around the body! They have a special code for differentiating location of sounds.

For example sounds come within 5 feet above the person's head, it might release an encoded sound as 'dit-dit-dit ' similar to the sound a Geiger

Counter emits when locating a treasure in the sand. On a different point in space, cells may make a sound such as 'dit-ditty-dit' firing in a different place. Locations are coded in this way. A single neuron can encode-sound from every direction in space. The cell will fire its signals in one pattern to the particular location vicinity, especially when unique sounds are released such as those found in a person's voice.

For example, a person with visual impairment would have language-sounds amplified in firing, to locate placement of frequency-sounds in his brain from the Temporal lobe. This research unveiled when the Apostles were on the mountain with Christ in scripture *2 Peter 1:17 "For he received from God the Father honor and glory, when there came such a "voice" to him from the excellent glory, and the "voice" which came from heaven we heard, when we were on the holy mountain."* The booming sounds of the voice of God had a residue encoded in the Apostle's brain when they heard Creator God's majestic voice above them.

I am a soprano opera singer, and play piano and synthesizer. My flying fingers hammered the ivory keys on a Grand Steinway piano, as I pitched a shrilling aria in the Ritz Carlton at Laguna Beach California. A Divine Opera rang, as liberty bells with "sound-waves" were ringing along the hallways of the hotel for a God-ordained span of time. People heard the cadenza sounds from various locations inside and outside the hotel, including the hotel swimming pools on the second level. The "Coloratura-Singing" struck like sweeping aural tornados. Troops of tourists, revolved like Saturn rings around me, as I released Aria-Opera-Sounds.

One person said they heard music flooding through hotel corridors, and was trying to figure out where it was coming from! People's brains encoded a surround-sound in their neuron's-circuitry thru the musical opera sounds. Recently I wrote, recorded and sang a wedding opera spontaneously. First I would hear the sounds, and then I would put them into written form as music later. I proceeded to play synthesizer keys and sing opera music to accompany the music fully aware that I was opening new pathways of thinking because I had never heard these new shrilling high notes before.

At first when you hear a soprano singing beautiful lyrics of "The Sun, Whose Rays" from Mikado Opera, you may not recognize the sounds. However after repetition of hearing the euphoric sounds, the synapses in your mind will adhere to the sound and you will recognize it. So as the aging process progresses, our Hippocampus subsides in its role of learning new things yet repetition will still cause our minds to cling to old memories.

A neurotransmitter is released in the Amygdala; the part of the brain that stores "emotional memories" to draw calcium into the cell. This simultaneously prompts special proteins into the nucleus that switches on the genes that routes proteins to synapse when the music plays. The synapse comes alive and cells fire, "voila" a charged memory occurs.

You're dreaming of "A White Christmas" again! Memories of gold may be panned from experience but they are rinsed in the brain's chemicals that effect emotions. For instance, I love butterflies. They indicate spring has bloomed, like a charging calf let out of the stall! Through my front door for a split second, suddenly a golden Monarch butterfly fluttered into my house as if he found a giant-sunflower. Or another occasion a hawk "zoomed" down my sister's house chimney, in hot pursuit of a frenzied parrot, like a kamikaze bomber!

A few years ago I was sailing off Catalina islands when a Flying Fish leapt out of the Caribbean-blue Ocean and torpedoed onto our 27 foot sailboat, simultaneously a huge shooting star exploded across an ebony sky. These are indescribable gifts of memory to cherish in our brain's vault of treasures. My grandmother was 90 years old when she passed away, and possessed a razor sharp memory, gifting generations far beyond her grandchildren!

We asked Grandma Eloise her secret of intelligence. She said, "I read, and engage in crossword puzzles and go out every day to shop or walk on excursions." She set high priorities on being a lifelong learner, and limited her television intake to a minimal amount of time about an hour or so as well as practiced a love of daily devotion to prayer. I believe this was the catalyst that enriched and filled her life with a stellar memory!

CHAPTER 14

YOUNG AND OLD BRAINS

Genesis 1:27 states: "So God created man in his own image, in the image of God He created them." At the dawn of creation beginning in the Garden of Eden, Adam and Eve had all their mind faculties blossoming into full bloom. When a baby is forming in its mother's womb there are estimated 200 billion-brain-cells, in their full anatomical operation. At twenty weeks the cells begin to self-assassinate each other reducing their capacity by half. The facts state that now humans only uses 10% of the capacity of their brain. Compare this to Adam and Eve's limitless abilities and their cognitive functions were so alive and sophisticated, they were able to communicate directly with Creator God and the animals. That is why it was possible for Eve to converse with the serpent. Contemplate for a moment the manifestation of divine communication in operation.

Considering the facts stated above that at twenty weeks the neurons in a baby's brain are destroyed – could it be that a baby in the womb has perfect knowledge prior to the self-assassination process of its neurons but, due to the impact of the Adamic nature, once born the enlightened state of the baby's brain begins to disintegrate. What would have been perfect in?

95

Knowledge is now usurped by the corrupt nature of man and brings about a 90% loss of functioning brain parts. This loss of superior cognitive capacity was determined by Creator God when he banished Adam and Eve from the abode of paradise. *Genesis 3:22 states "And the Lord God said, "Behold the man has become as one of us, to know good and evil and now, lest he put forth his hand, and take also of the tree of life and eat and live forever…- therefore the Lord God sent him out of the Garden of Eden."*

During the process of formation of a baby's brain, it organizes itself into forty different mapping roads which govern vision, language, muscle, movement and hearing. The brain can be compared to an amazing super sponge and is most absorbent from birth through to the age of 12. Thus the brain can reorganize itself during these early years, when connections between brain cells are made at an enormous rate. Information flows through 'windows' that open during the first three years as streams form together to make rivers of thought processes.

These processes house the foundations that form the structures for the characteristics of thinking, language, vision, attitudes, and aptitudes. When the 'windows' close much of the fundamental architecture of the brain is complete. Compare this to individual Lego pieces being snapped into place to form a complete model. A shaping process occurs between the ages of 2 through 4, and continues with the mapping system of organizing completed by age 12. The dynamics of the final result are influenced by genetics and environment.

The brain has the ability to organize itself in many positive ways – such as learning to play a violin or mastering calculus, creating visual art, dance, or drama. However, when the brain wires itself in negative ways learning disabilities result like obsessive compulsiveness, ADD, erratic behavior and other cognitive problems arise. Reorganization of our brain with more efficient circuits is necessary to replace inappropriate thinking. When we embrace new purposes, plans, blueprints, ideas and strategies to organize our brain into fresh ways of thinking - shifting the brain's paradigm mindset from an old routine to the new – we actually develop new brain trees capable of ground breaking thought processes.

Prior research has shown that ageing was thought to cause an irreversible downward spiral into mental befuddlement. However, new research shows this theory is little more than a self fulfilling prophecy. Common ageing symptoms in the brain are now seen as being directly related to brain disuse. According to Dr. Marvin Albert at Harvard's department of Neurology, the old school of thought taught there was a loss of nerve-cells every day – something close to a million – but clear evidence now shows there is not. When there is active participation with new activities the fact is that new ideas and sounds break old cycles.

The key to increasing brain cell ability and/or reversing the decline of brain cells was active involvement in mentally stimulating activities. Researchers did an experiment on two age groups – 25 year olds and 70 year olds.

The 25 year old age group of volunteers were injected with a rainbow display of radioactive sugar in their blood, and given a ½ hour memory test. Their Frontal lobes glowed bright yellow and red with computer imaging as the sugar was consumed and transferred into energy by billions of brain cells busily processing memory. Indeed this result was expected, but the surprise came when the Frontal lobe of a seventy year old shone just as brilliantly as the 25 year olds.

In 1962 there was a historical space launch that involved the Astronaut John Glenn. He was part of the team that flew in the Mercury capsule "Friendship 7." John was one of the seven Astronauts to first orbit the earth three times. Later in life he changed careers and went into politics to become a United States Senator. Then the summons came for John to go up again on a second voyage into outer space in a space shuttle at the age of 77. He had all his mental faculties glowing and functioning as brilliantly as they were during his youthful years. John made the news as a national hero!

My father was part of a group of men that lead a team to design the guidance systems for the astronauts that went to the moon in July 20 1969, with Neil Armstrong. As a result our childhood was spent associating with the children of families of astronauts that were part of the Houston swim team. We would occasionally be driven by John's daughter, Lynn Glenn, to the swim team-diving meets. On one occasion, we were in a car accident. We were rushed to the hospital, and John Glenn came to visit his daughter and us. This experience was thrilling; we forgot our minor bumps and bruises!

The amazing fact is that John Glenn, a former Astronaut and U.S. Senator could be so healthy with fully functioning Mental faculties at 77 years old! When a lot of people are mentally and physically spiraling downwards in the same age group, here was John, in his golden years a brilliant shooting star.

In *Joshua 14:10-11 Caleb instructed Joshua, in his older years, to possess a mountain located in Kadesh Barnea "Now then, just as the Lord promised, he has kept me alive for 45 years since the time he said to Moses, while Israel wandered in the wilderness. So here I am today, eighty five years old. I am still as strong today... as the day Moses sent me out; I'm just as vigorous to go out to battle."*

Changing your thoughts positively produces changes in your brain and possibly in your DNA. According to neuroscientists, we think up to 60,000 thoughts a day! New York Times best selling authors, Marci Shimoff and Carol Kline wrote a book entitled "Happy for No Reason". In it they refer to strengthening the "Pillars of the Mind". Of the 60,000 thoughts a day that traverse through our brain surprisingly eighty percent of those thoughts are negative. What is even more startling is that ninety-five percent of those thoughts are the same as yesterday. One's mind can be likened to an IPod or CD player stuck on the repeat function! God's word has a different perspective on this in *Lamentations 3: 21-23 "I recall to my mind and therefore I have hope because of the Lord's great love, we are not consumed. For his compassion never fails. They are new every morning."*

Creator God's thoughts towards us are refreshingly new and full of promise and hope. The fact is your thoughts are not always rooted in the truth. Studies have shown that brain waves are different in Alpha waves of happy people compared to the activity of unhappy people. These are key thought processes to helping to establish a young mindset by integrating an exercise of new habits as follow s:

- Question-your-thoughts
- Go beyond the thought and let go
- Incline your thoughts towards joy
- Is it a truthful thought?
- Do you know for certain it is a truthful thought?
- How do you react when you believe that questionable thought?
- What would your brain be thinking without that questionable thought?

When you exercise these new habits you cycle into a new tipping point in your mind. You will be astounded by how quickly you are able to cultivate fresh levels of creativity and lead your thoughts instead of them leading you! *Ephesians 4:8 & 3:20 states: "Therefore it is said, when he {Jesus} ascended on high, He led captivity captive and bestowed gifts."* God led captivity captive, therefore we have that same power in leading "captivity thoughts captive."

CHAPTER 15

CEO EXECUTIVE MIND

The brain contains unfathomable oceans of deep thinking that we can immerse ourselves in on a daily routine which is compared to a 9-5 job. We can streamline our thought processes into detailed higher levels of reasoning. Our brain's Frontal lobe is known as the multi tasking, reasoning area, judgment, and problem solving. You can think clearly, focused, and stable for the days ahead with Executive CEO Functions. For example when you board an airplane and suddenly they change your ticket to first-class-seating, you're swiftly upgraded into a CEO executive first class business world.

Take any area of work such as student, housewife, plumber, machinist, computer analyst etc. If an individual has a frenzied mindset and delves into thought processes of negative-erratic-emotions they soon fall into a Fight-or-Flight pattern trying to find a fire escape door. This is where we can lose our executive CEO functional thinking, and then our decisions no longer come from a solution-oriented-perspective, but from an emotional reaction based on ego or physical survival. Decisions are then made for what is good for the individual at that moment, rather than the ultimate good for everyone involved affected by their decision.

It becomes a selfish oriented, look out for number one, me first attitude. A fine example of this is CEO Paul Levy of Beth Israel hospital of 8000 employees. In the recession he had an idea where he decided to protect the lower wage earners. Paul said "The rest of us will have to sacrifice salary increases, bonuses and benefits." He felt sheer power rush over him like a wave. The employee consensus agreed none wanted the workers laid off. The employees realized everybody is in the same boat, and that their boat doesn't rise because someone else's sinks. Paul's radical-revolutionary-ideas noted that the "E in CEO means Empathy". King David also addressed the CEO of the universe in *1 Chronicles 29: 11-12 "Yours O Lord is the greatness and the power and the glory and the majesty and the splendor, for everything in heaven and earth is yours. Yours, O Lord, is the kingdom; you are exalted as "Head" over all."* So our CEO Creator God declares a dialogue of reasoning in our minds through prayer to bring solutions to individual's problems.

When our CEO functions are operating properly we will have creative problem thinking. But if our thoughts are confused or fragmented, we'll move into an option known as "Emotionally-Driven-Reaction" thinking. The brain knows who we are, but our immune system knows who we are not. Together, trillions of cells build a mosaic memory for the body and brain they defend. This reminds me of a King and his mighty army defending his castle walls; in this case the castle is our brain. A troubled mind can trigger hormones that may subdue an immune system. We usually think of choice as freeing, widening one's scope, but with every choice there is a world of competing options, however rich.

The brain has a bullet-proof-vested interest in protecting itself so our world will feel solid, safe, and predictable, not a careening-roller-coaster, where one clings for dear life.

A few years ago my brother Bill had a speed boat for fishing, one summer we were jetting up the river, and cruised by these buoys where we dropped anchor in the middle of the river. Suddenly a swarming black cloud of wasps hovered over us and began to attack and sting us. So my brother slammed the engine-throttle in reverse, jamming his boat gears where we could only travel backwards, cruising up the river. Whew we escaped the "Hoards-of-wasps"! Up and down the river canyons howling boaters passed us by laughing at us, as we putted along, floating in reverse past them. Our flight-or-fight! Fleeing syndrome evolved, as we were in dire straights.

When in this Fight or Flight mode your instincts summon strong-survival-emotions and your creative solutions-oriented thinking is not available. You will react out of survival emotions of fear, anxiety or anger. When your right foot presses on the gas, while your left foot stomps on the brake, you are in a standstill as an engine fights the brakes for control. God has elaborated on this and gives us the antidote in *2 Timothy 1:7* *"For God has not given us the spirit of fear; but of power, and of love, and integrated sound-mind."* This power, love and sound mind multiplies creative solutions, and key thought processes. A lot of leafing out and pruning goes into the DNA which sculpts our minds. Since each brain knows what topiary to sculpt, to strategize and plan abstractly.

Creator God has a pattern of a kingdom mindset that he's called us to found in *Deuteronomy 28:13"And the Lord shall make you the "Head" and not the tail; and you shall be above only, and you shall not be beneath."* There is an array of new tools to increase our brain ranks. As we age the crystallized abilities remain stable, but our so-called fluid-abilities decline. These have to do with planning, and multi-tasking.

So a study was done using a video game called the rise-of-the-nations. 40 people were recruited in the 60 year old category, and did role playing where you build your own empire. The game-player had to build cities, feed and employ their people and maintain an adequate military and expand their territory. Gamers were startled that by increasing spatial and visual exercises in their mind it actually assisted in learning new skills.

CHAPTER 16

GAMMA RAYS ON THE BRAIN

NASA Space Center launched into outer space on June 10, 2008 the first 700 million dollar Fermi-Ray-Gamma telescope. This will be used to investigate our universe, stars, giant-super-novas, black holes, and pulsars. Even the milky-way surrounding halo, is believed to be composed of Gamma Rays. Astrophysicists estimate Gamma Rays to be the highest form of light and energy, which is astonishingly invincible, since it makes up a massive portion of our universe. The brightest events ever occurred in our universe exploded from powerful Gamma Rays.

Intense flashes of Gamma Rays can release within seconds the same amount of energy, that our sun will put out over it's 10 billion year-lifetime. In a few major solar flare bursts of the sun produces Gamma Rays with energies up to one million electron volts! "Wow, now that's power!" The Fermi Gamma Ray telescope observed a particular pulsar sweeping over the earth like a lighthouse. These beams of Gamma Rays from the pulsar rain past the earth roughly three times a minute.

Pulsars are fast spinning Neutron stars left over after massive stars have exploded as a Supernova Gamma Rays have no mass and no electrical charge, just pure-electromagnetic-energy. Because of their high energy they can travel at the speed of light, they also have 50 million times the energy of visible light. These turbo charged cosmic blow torches release jets that set our universe ablaze. They can pass through lead, or human tissue. Sounds like super man with his penetrating x-ray vision. The domain of the high energy light of gamma rays spans about 7-orders-of-magnitude in frequency wavelengths, above the energy of frequency X-Rays light source. Fortunately Neuroscience has captured these powerful energy sources, Gamma Rays, to help with over a half million people with brain ailments.

A brain surgery treatment called "Gamma Knife" surgery is a state-of-the-art radiation laser. Neurosurgeons state "that no knife, or incision, is used but it targets and destroys brain tumors." It penetrates through the head, and is virtually painless with one night in the hospital. By focusing on one single point in the brain the radiation is harnessed and can virtually wipe out tumors, brain lesions, penetrating through every wall in the brain without hurting anything else in its path.

The interesting factor is Jesus said "I am the light of the world" with the algorithm in Creation in the word he declared *Genesis 1:3-4 "Let there be light," and there was light. And God saw that the light was good."*

The great source of God's explosive-power that holds the universe together is the Gamma Rays that move at the speed of light and are the highest form of energy. It wouldn't surprise me if Christ moved through walls when he appeared to the disciples, after his resurrection with the Gamma-Rays-Light-Source in his glorified-body. Dr. Daniel Amen has pioneered a "Spect" for the brain which stands for a unique process called "Single-Photon-Emission-Computer-Tomography".

This is nuclear medicine that allows visualization of brain-blood-flow, and metabolism, in a radioactive isotope. This is attached to a substance called Ceretac that is easily absorbed in brain cells. A small portion is injected into a patient's veins, and travels through the bloodstream, locking into brain cells.

Energy is released in the form of "Gamma rays" that strike as beacons of light signaling the compound substance in the brain. This reminds me of the TV show Star Trek, and the laser that beamed up Scotty, a character who was transported through galaxies. Interestingly people do not have allergic reactions to Spect studies. There are Special crystals in the Spect "Gamma" camera that detect these beacons of light as the camera revolves around patients head. About 10 million Gamma Rays strike the crystals during a typical scan.

A supercomputer translates this pertinent information into blood/flow maps and 3 dimensional images of the brain, which identify patterns of brain activity correlating with healthy brain functions. This amazing supercomputer helps those that are associated with psychiatric and Neurological illnesses, with this great breakthrough for patients who might be experiencing ailments in heart, cancer, thyroid, and mind etc.

An example of this is when a real estate inspector walks through a house assessing any damage, shining his flashlight into the structural foundation, or hidden cubby holes. Our Creator God shines as a beacon of light within us in *2 Corinthians: 4:6 "For God, who said, let light shine out of darkness, made his light shine in our hearts to give us the light of the knowledge of God in the face of Christ."* To me this is a tremendous revelatory illumination of his Creative spirit.

CHAPTER 17

MIND HACKING

A team of Neuroscientists developed an intrusive brain reading surveillance technology. This powerful technique allows them to look deep inside a person's brain and read their intentions before they act. The device is a scanner that is called the Tera-Hertz laser that looks around the brain to read a person's thoughts. The Tera-Hertz laser moves in wavelengths on the electromagnetic spectrum. It penetrates through bricks and human skulls. Like a shining torch, peering around for the writing on the wall in an individual's mind. Shouldn't your thoughts remain your own personal business? This would allow your mind to become a communal pool, rather than individual possession, which is an invasive method of accessing minds through mind control through an outside laser.

The Tera-Hertz lasers can also scan through people's clothes to see any weaponry, in hopes of catching a criminal or terrorist. The Heathrow and Luton airports in London have used this technique. It is none other than what a computer hacker does except in this case in the form of "Mind Hacking" with this trained light. By using transferred radioactive energy with light sources and powerful magnets it absorbs moisture from human tissue which disentangles thousands of neurons and encoded signals from the complex circuitry of the brain. Global Research reports that Tera-Hertz lasers are small enough to fit into portable devices. I think of Creator God as a Tera-Hertz laser because in *Hebrews 4:13 "Neither is there any creature that is not manifest in his sight: but all things are naked and opened unto the eyes of him with whom we have to do."*

All human thoughts and reasoning are read by Creator God as described in *Proverbs 20: 27 "The spirit of a man is the candle of the Lord, searching all the inward parts of the belly."* In the early sixties there was a buzz in the news that the Soviet Research was masterminding a brainchild in a class known a "Brainwashing Reorientation." The idea was to mass-zapp uninformed people groups to invoke them into newly oriented ideological ideas. They sought to control the minds of the masses. We are on the threshold of having data processors attached to the human body and mind. *2 Thessalonians: 2:2 "That you be not shaken in mind or be troubled neither by spirit nor by word nor by letter as from us, as that day of Christ is at hand."*

Four black boxes that are the size of a refrigerator in Lausanne Switzerland are known as the "Blue Brain." A supercomputer programmed to act like a real-neuron in a real-brain! Talk about a cloning procedure. The computer replicates with shocking precision, the cellular events unfolding inside a mind, this is the 1st model of the brain that's been built from the bottom up.

The mind has been revealed as a Byzantine machine. The Blue-Brain is connected to another computer screen that is filled with digitally rendered, virtual neuron tree branches. In full view their appearance quickly morph's into an expansive canopy, as dense as the Bavarian Black Forest. In your brain dendrites channel out to each other like branches grappling for the sunlight, synapse-by-synapse striking each other as a lit fuse, ready to ignite the dynamite neuron cell. Like 2 sparklers on the 4th of July spurting and sparking. Lay an unlit one against a lit one and a burst of fire and sparks ignite. Thus all the cells chatter their vital thoughts or memories, as a cacophony of firing neurons that communicate with each other. A reference in the word where Jesus says in *John 15:5 "I am the vine, you are the branches, he that abides in me, and I in him, the same brings forth much fruit."*

There is a unique silicone graphics supercomputer that operates in a 3-d film. This computer can voyage into the deep spaces of the mind. By striking a few keys from the Blue Danube by Strauss the computer will give a vivid picture image of an inter-neuron with spindly limbs reaching through the air. It shows thousands of colorful cells a journeying into the center of a mysterious brain. Like a movie experience viewing reality in details. No matter how much we know about our neurons, we still won't be able to understand how a group of molecular Ions, are equipped to become a Technicolor Cinema of consciousness on a computer screen. There are over 100,000 chemical reactions going on in your brain every second as a radio transmitter sending out communication systems in various

Electrical wave signals. Even if we could put our brains on autopilot we still have the capacity to rewire the hard-wiring thoughts into a new world, by retraining our brain in something new! Written in the book of *Ephesians 4:15, 23 "But speaking the truth in love may grow up into him in all things, which is the head even Christ. And be renewed in the spirit of your mind."* As we regenerate thoughts, we can excavate into the deep recesses of a furrowing-diamond-mine, I call it a diamond mind journey. The indication of this is found in the word of *James 1:16 "For every good and perfect gift comes down from above from the Father of lights."* We categorize and weave the needlework of neuron tapestries together and establish the Big-Picture. This provides us with a daily-organizer-planner. If indeed we move in our mind's Matriarchal thoughts then we can synergize as described in this passage of *Amos 3:3 "How can two walk together except they are agreed"?* So when our thoughts are aligned in agreement with Creator God then we are entrusted to walk through freedoms door. *2 Corinthians 3:17 "Because where the spirit of the Lord is there is liberty."* So we can be crowned with "liberty-thoughts" and break free from old ways that would keep us stagnated. By being your own Brain Coach you shift mental disciplines into a "Catalyst-Change-Agent." I use to be a hairdresser in a salon, and we used various chemicals as catalyst's to change an individual's hair color. Once the catalysts were released into the hair shaft they would alternate its structure and highlights, this would bring the client's hair color image into a new shade.

CHAPTER 18

CHANGING MIND'S EYE

Diane Ackerman had gone to the ornamental rose garden in the Rodin museum in Paris to record her comments about Rodin's art. She leaned over close to his statue, the Thinker and waited for a sound man to stop a ferret from scrambling across a wire. Amazingly our brain-waves can have a wild side while navigating us through rushing thoughts. Even when nothing is happening, our mind the "Thinker," sends slow, irregular spikes down the neuron's axon at a rate, between 1 and 5 hertz.

This steady flow of neuron action keeps them alert and ready to fire at breakneck speed. Firing faster, at 50-100 hertz or so, and for short increments it can rapid fire a hundred times faster. A flood of testosterone in the male brain causes brain wiring to take place. A flood of Estrogen promotes female brains to wire. Her brain contains a larger Corpus Callosum, which is the bridge between the right and left brain hemispheres. This is why women tend to use both sides of the brain. Emily Dickinson wrote in her poetry about the brain being wider than the sky.

The Brain – is wider than the Sky –
For – put them side by side –
The one the other will contain
With ease – and You – beside –

The Brain is deeper than the sea –
For – hold them – Blue to Blue –
The one the other will absorb –
As Sponges – Buckets – do –

The Brain is just the weight of God –
For – Heft them – Pound for Pound –
And they will differ – if they do –
As Syllable from Sound –

Emily's mind transported her words as flying arrows into a bull-eye at top speed. Likewise the brilliant trail blazer, Einstein, followed suit as he quoted ***"The world as he sees it is the most beautiful experience we can have is the mysterious."*** It is the labyrinthine emotion which stands at the cradle of true art and science." According to Diane Ackerman's writings, she states:

    "Our minds flicker like candlelight faster than we realize fed by a steady glow, if we count the moods in a day 5-45 but who can count the moods in a minute, where the thinnest membranes seal us in, which gives pastures to roaming cellular factories, neurons that fizzy with electricity."

There are many different dimensions to the thought processes. The following are some of the categories. Lightening speed reflexes, accelerated thought, focused analytical, daydreaming, prayer-meditation, intuitive thought, and slow meandering thought. With keys to a gateway of insight that roams our 7 seas of the unconscious like a submarine, whose wake sometimes becomes visible to the ever changing mind's eye. Your senses are able to refine and sharpen the focus of your mind, known as top down and bottom up concept. In our brain we all have a natural blind spot. Our eyes try to fill in the information. The top down and bottom up concept will help memory problems in your attention span. This trains the eye so you won't ignore necessary information.

For example you and your friend Jeanne are eating in the New York Cheesecake Factory. While the buzz of everyone's noisy conversation is distracting you from dialoging with your friend. This scenario becomes increasingly difficult for you to focus on your conversation with Jeanne, when in a crowd or large group. Otherwise illustrated in a zoom lens on a camera which magnifies a particular field of imagery so that the mind's thinking is connected with your ear gates to broaden your attention span. This focus will empower you to zoom in on your friend's conversation. This will simplify and enable your mind's top-down concept to clarify and obtain understanding for your friend's conversation, instead of all the distracting noise in the restaurant. Otherwise these attention signals flow in opposite directions.

So you're tuned into your friend Jeanne's stimulating conversation, especially as you splurge on that blueberry extra cheesecake. Cell Press released a news flash that demonstrated that researchers have for the first time revealed that a blind man navigated an obstacle course flawlessly after brain damage. He was left with no awareness or the ability to see and no activity in his visual cortex, which processes sight. This reveals the importance of alternative routes in our brain, which are active in those who have suffered severe brain damage to the visual cortex. For example, you were chugging along a road in your car and suddenly a big detour sign pops up so you have to turn onto an alternate road and travel a different way to your destination.

An unusual patient known as TN was left blind in his visual cortex, in both right and left hemispheres of his brain following consecutive strokes. TN has what is called Blind-Sight the ability to navigate and detect things in the environment, without being aware of seeing them. For instance, he responds to facial expressions of others, as indicated in brain regions of emotional expressions of fear, anger, and joy.

He nevertheless is totally blind. Researchers constructed an obstacle course of boxes and chairs and asked him to cross thru. Astonishingly, they reported he navigated thru the maze. This demonstration shows that alternative visual paths are available in our mind. *Ephesians 1:18 The eyes of your understanding be enlightened.* Allowing people to orient themselves and rapidly detect obstacles in their environment without any conscious attention or experience of seeing them. All the time we are using hidden resources of our brain and

Doing things we think we are unable to do. The Spirit man in each one of us causes us to walk by faith not by sight, as reported in the Elijah List. *Job 32:8 "But there is a spirit in man, and the breath of the Almighty gives them understanding."* Primarily when things aren't going right in one area in our lives, there are other hidden pathways along life's journey that transcends another.

Our brain has its own natural "Happy Hormones" called Endorphins that act as jubilant joy centers on the brain and help us cope with pain or emotional turmoil. If these brain hormones become depleted, there might be an occurrence of gloomy depressive mood swings. Regarding an imbalance of these various hormones it creates malfunctions such as poor memory, insomnia, panic attacks and allergies. According to hormone control it influences so many aspects of human behavior to make you mentally competent, and functioning with zeal for life.

Neuroendocrinology is the study of the complex release of hormones in the endocrine system as it interacts with the mind and nerves. These different hormones that are released and pass through the pituitary glands are controlled by the Hypothalmus. The fluctuating of hormones are the root issues that causes an adverse reaction to those diagnosed with Bi-Polar, Manic-Depressant, and Schizophrenia, symptoms which have a chemical basis.

An example of Schizophrenia is found in the story of King Nebuchadnezzar the great Emperor of Babylon. His mind Was changed into the mind of a cow where he grazed on grass in pastures as an animal for seven years. After a long seven years his reasoning, understanding and intelligence was restored back into his brain. As stated in ***Daniel: 4:36-37 "At the same time, I Nebuchadnezzar raised my eyes toward heaven, and my sanity was restored. Then I praised the Most High; I honored and glorified him who lives forever. His dominion is an eternal dominion; his kingdom endures from generation to generation."***

When blood levels were measured with the intake of vitamin B-12 between 60-80 year olds, they found that they were mentally and physically, sharper in their awareness of their livelihood. They found those with low B-12 were six times more likely to have brain-atrophy which is brain shrinkage that is linked with Alzheimer's and other mental decline. Doctors say by simply adjusting our diets to include more B-12 we can prevent this brain shrinkage and save our memory. There was an accelerated mental decline over young people minds in comparison to the older people group who took an increased B-12 vitamin over a 10 year study. I use to take Karate and achieved a Gold belt not quite good enough for a Black belt but I remember our Grand Master "Jimmy Woo" who was well advanced into his upper 70's. He was amazingly agile and in his peak performance could outdo a 20 year old. He confided that he supplemented his mind with "Golden-Ginseng" as well as discipline and action.

CHAPTER 19

BRAIN WISDOM

When your brain feels power, you are enhanced with increased mental aptitudes, in a world that at times is teetering with chaos. Your mind can be stable, and focused. This is found in *Isaiah 33:6 "And wisdom and knowledge shall be the stability of your time, wealth of salvation and fear of the Lord his treasure."* Wisdom comes naturally to older people. Your neurons are constructed to contain wisdom, with age. New brain trees or dendrites have made a new circle of connections, a neuron social club all chattering together, as a thundering stampede of mustang horse hooves rumbling across the land.

A young person's brain doesn't yet maintain this multiplied network of billions of cell connections that Creator God gives to you as you age. As in *Proverbs 8:1 "Does not wisdom cry out and understanding put forth her voice."* But not everyone who ages achieves brain wisdom. Some people gobble up wasteful facts and if new knowledge doesn't bring the way of aligned truth, it's simply trivia not knowledge. Since our mind has plasticity, it can renew itself, or retrain the brain. As so eloquently portrayed in *Romans 12:2 "And be not conformed to this world: but be transformed by the renewing of your mind that you may prove what is that good and acceptable, and perfect will of Creator God.*" Controlling stress makes you wiser since your brain has infinite potential for increased learning and happiness. You can pioneer a new pathway by using your imagination instead of movies or video games. Consider a sailor, sailing through uncharted Caribbean blue waters, lined with oyster beds of glistening-hidden-pearls. Likewise we contain a vast treasury where our creative thoughts maintain exclusive blueprints, and copyrights to your mind's progress, in the brain.

Exercise with proper diet revs up cognitive power. This will assist in keeping a clean sweep from a barrage of cluttered thoughts from crowding in and breaking concentration. I was once in Mexico and sat across from a hillside that was covered in a worn maze of trails. I couldn't imagine what would have made these well worn paths. Suddenly a rabbit runs all around this maze and then off he goes. This is where we get the term "rabbit trails" Instead of going from point A to point B you run in circles before getting to your destination. We cannot become a life-long master brain-builder with passive cognitive skills.

On the 2nd leg of life I entered into the weight gym to rebuild my muscles. This was not easy for me. What I really needed was a little bit of spinach and Popeye's strength. I was so intimidated by the Mr. and Miss Universe body builders all lined up with their perfectly sculpted physiques. A pulsing parade primed for muscle building marched by me. The competition looked more like Paul Bunyan and his Blue ox, or when David faced off Goliath, with his sling shot. I pulled myself together and changed the lever from 150 pounds to a mere jingling 15 pounds on the heavy-iron-robot-machine. While I tugged on my forlorn muscles to respond I repeated like the little train that proclaimed "I think I can, I think I can." An earthquake seemed to rattle through my bones and muscles, as I crackled and squeaked, thrusting into a new Universe, as when a jet breaks through the sound barrier.

But once I started exercising those limp unused muscles "Vroom" they stood into regimented place and the change was remarkable. So when we retrain our Brain-Gym into new patterns, we move against our opponent of mental befuddlement. Like a chess game player moving the King, Queen and Pawn's into new positions of rule on a chessboard. We'll stay three steps ahead of the game of learning in our neural-network. Use it or loose it.

A celebratory buzz arose from the Science group, making the occasion sound more like a birthday party or wedding than an Academic meeting. It was, in some ways, though, both wedding and birthday, as "Neuroethics" was birthed. The study of Neurology was married to Ethics. The study of "What is right and wrong, good and bad.

In summary it is about the treatment of perfection, the unwelcome invasion of manipulation and the privacy of our minds. Questions would arise like "What about "Botox for the brain," that would make people less shy, more honest, or more attractive with a nice sense of humor.

One incredible factor in your brain is that "Love conquers all"! The heartbeat will increase when someone's thoughts are in love. The simplest emotion of love invigorates our nervous system by simply being affiliated with it. This will strengthen your Amygdala, which contains all your emotionally stored memories. When someone is in love, they are more creative according to doctor Diamond. As expounded in *Corinthians 13:2 "And though I have the gift of prophecy, and understand all mysteries, and all knowledge, and though I have all faith, so that I could remove mountains, and have not love, I am nothing."* When the heartbeat of our thoughts is centered in the oasis of love the heart pumps blood to nourish the body. We can see how love is the key that embellishes our brain for wholeness.

CHAPTER 20

BRAINSTORMING
APPLES - CRUSHERS

A summary of garden fruit ideas comes from a tree of life mindset I call "Apples" which means, apply-personal-promise-learned-experienced-soaring. Imagine a crowd of people in a room, chanting out as a cheer-leading squad. "No limits, no boundaries, I see increase all around me stretch forth, break forth, and increase our mind's territory." Suddenly A Shofar Trumpet blast awakens the brainstorming session. A whirlwind of great ideas unfolds from engaging minds together, in this brainstorming session.

For example 75 different ideas from cultural groups with unique thoughts, are plopped on the brain-child table. These resemble more or less the image of a raging-river of ideas, or a group of sky-jumpers floating from the sky. I picture volumes of picture images evoking creativity with precision timing all the while being sure to banish naysayers and negative thinking. Good manners prefer one another, and fertilize each other's thoughts. Of course some ideas won't be good, some will be outright absurd. So you just eat the meat and spit out the bones.

So the antithesis of Apples is "Crushers" which is described as cancel-rushing-unusable-surfaced-hurtful-excuses-rebellion-selfishness. When you become a Crusher it manifests in the things you say like "I won't get that job promotion," or "You'll never amount to anything." When this happens you withdraw within a world of your own called selfishness and settle for a bottle of vodka. While wallowing in thoughts of self pity, excuses ring like a school-tardy-bell. Or you begin a diet and your thoughts echo "I won't lose that excessive fat so I'll eat a box of Famous Amos cookies." Suddenly your thoughts are ebbing with defeat and narcissism as a life-draining flow.

As Allen Faubion quotes "your brain tags memories of trauma." For example you might think "I'm no good" or have a root issue of abandonment. You can fill in the blank. So a "trauma lie" lodges in the memory banks of the brain. This area contains all of your emotional memories. So when I do Brain-Tuning discussed in an earlier chapter I dislodge this trauma-lie from your emotional memories so that you are free in your thoughts!

When you bless the Lord a promise of God's identity replaces the memory, and dislodges memories of brain trauma. Like cannon balls that are fired at "Crusher" negative thoughts, penned in *Psalm 103:1:4 "Bless the Lord Oh my soul and all that is within me bless his holy name. Who redeems your life from destruction and crowns you with loving-kindness and tender mercies."* A new crown of blessing rests upon your mind, which releases new truth of God's blessing on your life.

Sometimes Crusher thoughts pop up like popcorn. Suddenly your brain chemicals flood out and a charge zips through your brain. A giant sleeping Rip-Van-Winkle arises to pulsing thoughts, suddenly your body language changes with mood swings. You might encounter upset thoughts, stress or unkind words that attach to you like a dragging anchor in your limbic system.

There is a dynamic word that will help in scripture *2 Corinthians 10:5 "In as much as we refute arguments and theories and reasoning's and every proud and lofty thing that sets itself up against the true knowledge of God: and we lead every thought and purpose away captive into the obedience of Christ!"* These thoughts become detours and roadblocks and we can evict them from our minds vocabulary, and lock the door on "Crusher-thoughts!"

Mark George M.D. studied 10 women under 3 conditions when they were thinking happy thoughts, neutral thoughts and depressed thoughts. This reminds me of the popular movie with Robin Williams in Hook as Peter Pan where he couldn't fly until he found his "happy-thoughts." what Dr. George found was that the women with the "happy thoughts" had a cooling effect on the Limbic system.

During sad thoughts he noticed increased impressions deepening into their Limbic system. The blatant truth is your thoughts change your mental environment.

Every time you have a "Zenith" happy thought or kindness, or a gracious-thought your brain will release a chemical so your body responds positively. A fun game I loved to play as a kid was Candy Land. You move your game pieces through different sweets on the game-board obtaining a prize. When you think opposite of negative, anxious, muddied thoughts you'll find a honey-hive of Sweetness has evolved in your brain's activity.

You're not going to find the meaning of life hidden under a rock written by someone else. You'll only find it by giving meaning to life from inside yourself. Revealed in *Luke 17:20-21 "For the Kingdom of God doesn't come with "observation" but the Kingdom of God is within you"!* A Kingdom mindset uses signal-strengths and directed-thinking. Where is your mind's observation deck? Is it in a grandiose ship, like Noah's ark, full of everything to create a new world and make a world of difference! Or is it in the sea of drowning humanity awash in their sins and selfishness?

There was a New York Times best selling book by Spencer Johnson M.D. called "Who Moved My Cheese"? This is about 2 character mice called Hem and Haw, that lived in a maze and had cheddar cheese provided for them everyday. But one day the cheese was gone, so Haw the chief mouseling protested. "We need to venture into our maze to discover new cheese. We must take action to solve the problem instead of waiting for a solution to appear." So they scrambled about, finding new sources of cheese for themselves.

In our mental circumstances we don't have to wait and whine; but deal with it as Hem and Haw. The metaphor is "Who-Moved-My-Brain" which is a labyrinth maze of tunnel paths. Mazes engage the left brain when solving puzzles, as in a labyrinth. This allows a freedom in the right brain to reminisce or pray and is not designed to be difficult to navigate. Labyrinths are located at the city square downtown.

They have also been built into hospitals, gardens, and colleges, according to Daniel Pink in his book "Whole-New-Mind." He entered the circle as an escape labyrinth for his right brain as he continued to pray and walk, the left brain engages in logical progression of walking the path while the right brain is routed in a creative-mode-thinking, circling in the center.

In the center there are sayings etched in the walls such as" Create" "Faith" "Wisdom" "Believe." Creator God has his own labyrinth circle in *Isaiah: 40: "It is he who sits above the "circle" of the earth."* There are over 4000 of these labyrinths in the medical world. These were once dismissed as woo-woo suggestions from new-age crinkled-crack-jobs, but is now a "Commonwealth" to help people achieve mental relaxation.

When Lauren Arttress an Episcopal minister at Grace church traveled to Chartres Cathedral in France she found a forty foot diameter labyrinth etched into the floor of the church nave. Chairs that hadn't been used in 250 years covered the floors. Arttress removed the chairs and walked the labyrinth and imported the concept to America, a good example of her walk is revealed the scripture in *2 Corinthians 5:7 "For we walk by faith not by sight"*!

As Harvard professor Ellen Langer says too many stumble through life mindlessly, stuck in routines and unaware of their surroundings. So when we breakthrough that flimsy glassed bubble mindset, there is a new pathway to creativity. One of my favorite places to do prayer walks is in Pasadena Arboretum in California where they have huge exotic renowned flower gardens and fountains in bloom in the springtime. They have their own unique harem of spectacular peacocks that roam the expansive grounds. While they spread their Ocelli feather fans in defense of their paradise gardens.

REFERENCE NOTES ON CHAPTERS

Chapter 1   Einstein's Mysterious Brain
Jerusalem Hebrew University, Dr. Marion Diamond former
Head of UC Berkeley Science department, Albert Einstein,
{Nobel Prize-Winner} Photo-Electric-Effect, Princeton
Hospital, Pages 1-8

Chapter 2   Paradise Brains
Michael Moizen M.D. power plant, Jeff Victoroff M.D. author
of "Saving your Brain," Rob Stein's Washington Post Article
on "Deep-Brain-Stimulation" NKJV word, Pages 9-16

Chapter 3   Brains Internal Maps
Brain Trees Neuroscience Cell magazine, NKJV   Pages 17-22

Chapter 4   Brain tuning
The superhighway of aligning all key thought processes with
Creator Pages 23-34

Chapter 5   The Voyage of High Rise thinking
Medical records and diagrams, NKJV word Pages 35-40

Chapter 6   Brain Parts and Functions
Dr. Ward Bond PhD. From T.V. Show Nutritional Living,
Pulitzer-Prize-Winner Science writer Ronald Kotulak of his
book "Inside the Brain" Sheila Murray Bethel Ph.D. author of
New Breed Leader, Pages 41-49

Chapter 20   Brainstorming-Apples-and-Crushers
Dr. Mark George mental health specialist, Harvard Professor
Ellen Langer, Daniel Pink book author of "Whole New Mind"
Lauren Arttress, Spencer Johnson M.D. New York Times best
selling author of book "Who Moved My Cheese?" Allen
Faubion, mind and trauma healing, Pages 133-138

If you would like to receive or be trained in Brain Tuning
You can contact
Margaret Hardway

Via e-mail

Info@BrainGodtime.com

www.ingramcontent.com/pod-product-compliance
Lightning Source LLC
Chambersburg PA
CBHW031943190326
41519CB00007B/635